The Brain–Bladder Axis in Tissue Growth and Remodelling

IPEM–IOP Series in Physics and Engineering in Medicine and Biology

About the Series

The Series in Physics and Engineering in Medicine and Biology will allow IPEM to enhance its mission to 'advance physics and engineering applied to medicine and biology for the public good'.

Focusing on key areas including, but not limited to:

- clinical engineering
- diagnostic radiology
- informatics and computing
- magnetic resonance imaging
- nuclear medicine
- physiological measurement
- radiation protection
- radiotherapy
- rehabilitation engineering
- ultrasound and non-ionising radiation.

A number of IPEM–IOP titles are published as part of the EUTEMPE Network Series for Medical Physics Experts.

The Brain–Bladder Axis in Tissue Growth and Remodelling

Roustem N Miftahof and Christian J Cyron
Eißendorfer Straße 42 (M15), Technical University, Hamburg-Harburg 21073, Germany

IOP Publishing, Bristol, UK

ISBN 978-0-7503-3567-6 (ebook)
ISBN 978-0-7503-3565-2 (print)
ISBN 978-0-7503-3568-3 (myPrint)
ISBN 978-0-7503-3566-9 (mobi)

DOI 10.1088/978-0-7503-3567-6

Version: 20210701

IOP ebooks

British Library Cataloguing-in-Publication Data: A catalogue record for this book is available from the British Library.

Published by IOP Publishing, wholly owned by The Institute of Physics, London

IOP Publishing, Temple Circus, Temple Way, Bristol, BS1 6HG, UK

US Office: IOP Publishing, Inc., 190 North Independence Mall West, Suite 601, Philadelphia, PA 19106, USA

To all those who are always in our hearts

—Roustem N Miftahof

To my wife, Anna, and my daughter, Amelia

—Christian J Cyron

Contents

Preface

This book is intended as proof-of-concept as opposed to a pure research monograph. Its originality and novelty lies in its holistic approach to describe mathematically and simulate computationally growth and remodelling of the urinary bladder within a unified framework of the discernible and physiologically defined brain–bladder axis. At the time of writing (and this statement appears in almost every chapter), no precise morphological or quantitative experimental data was available on: the human cytoarchitecture of the brain–bladder axis, electro-physiological properties of cells and neurons of the spinal dorsal horn, periventricular nuclei, arcuate nucleus-median eminence complex, somatotrophs and insulin-growth factor-1 producing cells, the neurotransmitters' and hormones' cognate receptor dynamics, kinetic rates of biochemical reactions and concentrations of the relevant reactants involved. Therefore, most of the conjectures made in the construction of the model are based on deductions and guestimates with only a few confirmable facts. Despite all the imponderables, this is an endeavour to integrate a conceivable sequence of events and, thus, to demonstrate how modelling and simulation can benefit the community of (mechano)biologists to complement their limited experimental opportunities and natural intuition in disentangling the convoluted neuroendocrine regulatory mechanisms of soft tissue growth and remodelling. This book lights the way by providing a coherent description and explanation for intertwined intra-cellular pathways in terms of spatiotemporal, whole body tractable representations: *Anche se non è vero, è ben trovato* (even if it is not true, it is well founded). There is the sincere hope that followers will advance and improve the model to make it more accurate, more robust and more reliable.

In writing the 'simple' language of mathematics and biology was chosen, i.e., a reader familiar with algebraic, ordinary and partial differential equations and biological terminology will find the text comfortable to follow. On just a few occasions has it been necessary to use the tensor apparatus. Therefore, the prerequisites for the book are a familiarity with the basic principles of cell and molecular biology, biochemistry, differential equations and the mechanics of solids. The reader may need to consult textbooks on these subjects before proceeding to read the book. In terms of applications, the book could serve as supplementary material in computational systems biology and bioengineering classes, for research seminars at first-year graduate level, and as a companion for researchers and instructors of applied mathematics and biomedical professionals.

The book was written at the Institute of Continuum Mechanics and Material Sciences at the Technical University Hamburg-Harburg, and was supported by Deutsche Forschungsgemeinschaft, Germany under grant DFG 386349077, 'Mechanically controlled growth and adaptation processes in the urinary bladder'. We are grateful to the administration of these institutions for their support. We extend our thanks to W J Attwood who carefully read and edited the manuscript and the publisher IOP Pbl for publishing the book.

Finally, we find it true that ‘the writer should seek his reward in the pleasure of his work and in release from the burden of his thought; and, indifferent to aught else, care nothing for praise or censure, failure or success (W Somerset Maugham)’.

Prof. Dr-Tech. R N Miftahof
Prof. Dr-Ing. C J Cyron

Author biographies

Roustem N Miftahof

Roustem N Miftahof is an Emeritus Professor, Dr of Medicine and Applied Mathematics. Internationally acclaimed as a leading scientist in the field of systems computational biology and medicine, Professor Miftahof has authored and co-authored seven previous books in these subjects. His career has encompassed both academia and industry across Europe, the Far East, the Middle East and North America. He is currently a Research Professor at the Technical University of Hamburg-Harburg, Germany.

Christian J Cyron

Christian J Cyron is a Professor and Head of the Institute of Continuum and Material Mechanics at the Technical University of Hamburg-Harburg, Germany. Professor Cyron is a known scientist in the fields of multi-scale and multiphysics simulations, machine learning and data-driven methods. This is his first contribution to a book on the application of methods of numerical simulation to study mechanobiology of tissue growth and remodelling in a complex multi-scale biological system.

Acronyms

AC	adenylyl cyclase
ACh	acetylcholine
AD	adrenaline
ALS	acid labile subunit
AMPA	α-amino-3-hydroxy-5-methyl-4-isoxazole-propionic acid
AP	action potential
ArcN-ME	arcuate-median eminence
$ASIC_{2a/3}$	acid sensing ion channels
ATP	adenosine-5‘-triphosphate
BD	Brownian dynamics
BK_{Ca}	large Ca^{2+}-activated K^+ channel
cAMP	cyclic adenosine monophosphate
CNS	central nervous system
CX3CR	chemokine receptors
DAG	diacylglycerol
DRG	dorsal root ganglion
$EAAT_3$	excitatory amino acid transporter
ECM	extracellular matrix
EGF(R)	epidermal growth factor protein kinase and its receptor (R)
EPSP/IPSP	excitatory/inhibitory postsynaptic potential
ER	endoplasmic reticulum
ERK-1/2	extracellular signal-regulated kinase 1/2
G&R	growth and remodelling
GABA	γ-aminobutyric acid
$GABA_A$	γ-aminobutyric acid receptor type A
GAD	glutamate decarboxylase
$GAT_{1/2/3}$	transmembrane GABA transporters
GDP/DTP	guanosine diphosphate/guanosine triphosphate
GH-(R)	growth hormone and its receptor (R)
Ghr(R)	ghrelin and its receptor (R)
GHRH-(R)	growth hormone releasing hormone and its receptor (R)
Glu(R)	L-glutamate and its receptor (R)
Gly(R)	glycine and its receptor (R)
GlyDc	D-amino acid oxydase
GlyHt	glycine hydroxymethyl-transferase
GPCR	G-protein coupled receptor
GPT	glutamate pyruvate transaminase
GRN	gene regulatory network
HIF-1α	the activation of hypoxia induced factor-1α
IGF-1(R)	insulin-like growth factor-1 and its receptor (R)
IGFBP	insulin-like growth factor binding protein
IP_3	inositol-1,4,5-triphosphate
IRS-1	insulin receptor substrate
JAK	tyrosine Janus kinase
KA	kainite
K_{ATP}	ATP–sensitive K^+ channel
K_{dr}	fast-delayed rectifier K^+ channel

K_{ir}	inwardly rectifying K^+ channel
K_v	voltage-gated K^+ channel
LDCV	large dense-core vesicles/granules
MAPK	mitogen-activated protein kinase
mGluR	metabotropic glutamate receptor
MLCK	myosin light chain kinase
NFκB	nuclear factor κB pathway
NMDA	*N*-methyl-D-aspartate
NO	nitric oxide
NPY	neuropeptide Y
$p27^{Kip1}$	cyclin-dependent kinase inhibitor 1B
P2X(Y)	purinergic receptors
p75NTR	protease-activated neurotrophin
PACAP	pituitary adenylyl cyclase-activating polypeptide
PDE	phosphodiesterase
PeVN	periventricular nucleus
PI3K	phosphatydilinositol 3-kinase
PIP_2	inositide-4,5-biphosphate
Pk(...)	protein kinase (A, B, ...)
PLC	phospholipase C
RANTES	regulated on activation normal T-cell expressed and secreted
SDH	spinal dorsal horn
SK_{Ca}	small conductance K^+ channel
SMC	smooth muscle cell
SOCS	suppression of cytokine signalling
Som	somatotropic cell
SOS	Son of Sevenless
SP	substance P
SS	somatostatin
S6	ribosomal protein S6
S6k	ribosomal protein kinase
S6k1	ribosomal protein S6 kinase polypeptide 1
$sst_{1/2}$	somatostatin receptors type 1 and 2
STAT	signal transducers and activators of transcription
TRPC(...)	transient receptor potential channels
VGAT	GABA transporter
VGluT	glutamate transporter
VIP	vasoactive intestinal polypeptide
UB	urinary bladder
UBOO	urinary bladder outlet obstruction

Roustem N Miftahof and Christian J Cyron

Chapter 1

Anatomical and morphological preliminaries

I feel an indescribable ecstasy and delirium in melting, as it were, into the system of being, in identifying myself with the whole of nature …

— Jan J Rosseau

A concise overview provides information on both anatomical and morphological structure of the human urinary bladder and its functional relationships with the regulatory growth and remodelling centres in the brain. Emphasis falls on a description of electrophysiological characteristics of the cells, neurotransmitters and hormones involved and their relative receptor distribution.

1.1 Overview of the urinary bladder

The human urinary bladder (UB) is a hollow musculomembranous organ located deep in the pelvic cavity. Anatomically the organ is divided into three major parts: the apex, body and fundus (base), the latter consisting of the trigone and neck (figure 1.1). The fundus is imbedded in the prostate in males and in the musculofibrous tissue in females, and is intimately attached to the internal urinary sphincter through the neck. The fibrous fascia endopelvina provides an additional connection between the base, the pelvic wall and the rectum. The entire body of the organ is enclosed in the loose fatty tissue of the paravesicular fossa. The apex is covered by a thin stretchable peritoneum to form a series of folds—the false ligaments—which, however, do not bear any biomechanical significance. The parts of the bladder are interconnected by the anterior, posterior, superior, right and left lateral walls to form a smooth surface.

Histomorphologically, the wall of the human UB consists of four layers: the mucous (urothelium), submucous, muscular, and serous layer. Details related to its morphology can be found in many textbooks and research monographs (Elbadawi 1991, DeLancey *et al* 2002, Campbell-Walsh 2007). Thus, only a few aspects

doi:10.1088/978-0-7503-3567-6ch1

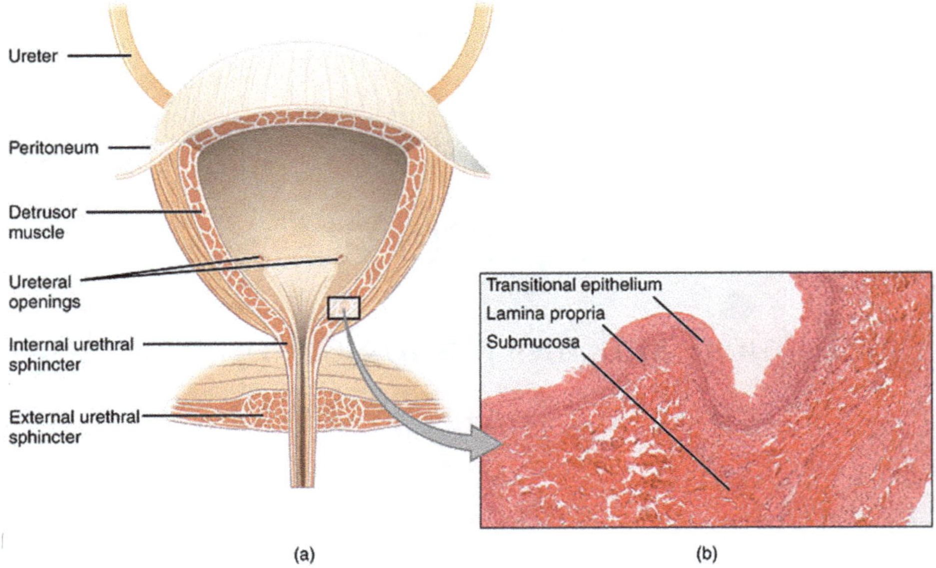

Figure 1.1. The human urinary bladder. Author: OpenStax College (CC BY 3.0).

relevant to the biomechanics of the organ are discussed here. The innermost urothelium (*tunica mucosa*) is made out of polyhedral shaped cells of stratified transitional epithelium including basal, intermediate and umbrella cells. The outer umbrella cell layer interfaces with the urine forming a primary barrier that includes a mucin/glycosaminoglycan layer which may prevent bacterial attachment and diffusion of urine components across the epithelium, as well as an apical plasma membrane with low permeability to urea and water. In addition, the tight junctions of umbrella cells form a close seal between adjacent cells and are comprised of multiple claudin species which regulate paracellular transport. The urothelium maintains the barrier even as the bladder undergoes cycles of filling and voiding. This accommodation can be explained by the ability of the highly wrinkled mucosal surface to unfold. The increase in the mucosal surface area results from the fusion of a population of subapical discoidal/fusiform vesicles with the apical plasma membrane of the umbrella cell layer (Apodaca *et al* 2007).

The submucous layer (*tunica submucosa*) contains a large number of collagen and elastin fibres, myofibroblasts and areolar tissue. Scanning electron microscopy studies have shown that flat, tape-like 2.1–6.2 μm wide bundles of collagen fibrils are interwoven to make a distinct network of the two-dimensional stroma (Murakumo *et al* 1995). Additionally, there are: (i) a loose network of twisted collagen strands 1.7–2.0 μm in diameter which are curled when the bladder is empty and straightened out when the organ is filled with urine, and (ii) a thin felt-like layer of dense collagen fibres running in all directions around the urothelium. Elastin fibres, however, are sparse throughout. Shaped like cords, 0.1–0.4 μm in diameter, they intertwine together to form a fine loose network around the blood vessels and submucosal muscle fascicles.

The muscular layer (*tunica muscularis*) is the most prominent comprising of three layers: internal, middle and external. Smooth muscle fibres of the inner and external layers run longitudinally from the fundus to the apex, while the muscle elements of the middle layer have a predominantly circumferential orientation. Together these layers form the detrusor muscle. The muscle fascicles and cells are firmly covered with collagen sheaths to form a regular honeycomb structure with 2–5 μm fenestration and 2–4 μm depth. Elastic fibres, 0.5–3.0 μm in diameter, are found primarily on the surface of the muscle fascicles. Entangled with each other, they are arranged transversely to the axis of the muscle fascicle. Sparse intramuscular elastic fibres are 0.1 μm thick and usually run parallel to the long axis of SMCs.

The serosa (*tunica adventitia*) is derived from the peritoneum. It contains homogenously distributed wavy collagen bundles of 2.0–6.2 μm thickness. These are piled up in a deep sheet (50–60 μm) with intercalated clusters of adipose cells. Ultramorphological and sonographic measurements of the entire wall thickness, h, of the human UB reveal: $h \approx 3.3 \pm 1.1$ mm with *tunicae mucosa et submucosa* comprising $\approx 1.4 \pm 0.2$ mm, *tunicae muscularis* $\approx 1.6 \pm 0.3$ mm, and *tunica adventitia* $\approx 0.3 \pm 0.1$ mm (Murakumo *et al* 1995, Hakenberg *et al* 2000). This remains relatively constant throughout the different regions of the organ.

The human UB, as described above, represents a multicomponent dynamic system. Optimal spatiotemporal arrangements among its anatomical and cellular/subcellular components guarantee the normal function of the organ. To sustain effectively its fundamental operation of the storage and voiding of urine throughout life, the UB, as any visceral structure, is subjected to continuous tissue growth and remodelling. Such vital adaptive processes are mediated by polymorphic neural circuits organised in meta-levels and located *in situ*, in the peripheral ganglia, spinal cord and central nervous system (CNS). Dynamic interactions among them are both moderated by diverse anatomically spaced non-neuronal and neuronal cell populations. Interconnected through polysynaptic projections into networks of afferent-efferent neuronal pathways, they form a unitary self-regulated closed system. It is highly sensitive to external/internal input signals and extremely versatile in its adjustable outputs.

Depending on the nature of pertubations, the system organises reflexes, and produces and releases chemically active substrates with up- or down-regulatory effects. Full comprehension of the sequence of events leading to such effects necessitates a detailed reconstruction and quantitative assessment, if possible, of the coordinated interplay among: morpho-functional sensory (first-order) neurons in the bladder; the dorsal root ganglia (DRG) at the level of S2-S4 and T11-L2 spinal segments; the intermediate secondary-sensory neurons (laminae I-II of the spinal dorsal horn) in the spinal cord; the periventricular nuclei (PeVN) and the arcuate-median eminence (ArcN-ME) complex in the hypothalamus; somatotrophs in the anterior pituitary gland in the brain, and hepatocytes in the liver, thereby instituting and evaluating the UB—brain axis.

The following discussion summarises the key features of the main components that give rise to physiological UB tissue growth and remodelling. Our current conceptualisation of neuronal cytoarchitecture and the mechanisms of information processing in the spinal cord and CNS is based principally on the results of investigations carried out on isolated and cultured neurons, neuron assemblies and anatomical regions of different animal species at various stages of the neuraxis and under various experimental conditions. There remains considerable disagreement between investigators on the cell types present and on the terminology used. It is instructive, therefore, to bear this fact in mind when interpreting and applying any findings to the human.

Results of this investigation and their applications are difficult to overestimate. They would offer quantitative understanding of pathophysiological mechanisms, develop new improved diagnostic modalities, and lead to effective personalised treatment of patients with UB dysfunction (Lopez Pereira *et al* 2002, Ansari *et al* 2010).

1.2 Detrusor muscle

The morphostructural unit of the detrusor is the smooth muscle cell—the myocyte. It has a characteristic spindle-like shape measuring ~100–300 μm in length and ~5–6 μm in diameter. Its cytoplasm contains a centrally located nucleus, intracellular thin α and β-actin (~6 nm), intermediate, mainly desmin (~10 nm) (Malmqvist *et al* 1991a, 1991b), and thick (~20–25 nm) filaments, mitochondria and fairly sparse elements of the sarcoplasmic reticulum. Thin α and β-actin filaments are arranged into a lattice attached to the cell membrane at the sites of dense bands (plaques). They guarantee the integrity, strength and high degree of deformability of the bladder wall and provide binding sites for myosin thick filaments (Martin *et al* 2007).

Regularly spaced dense bands are comprised of multifunctional proteins: integrins, desmin, vincullin, tensin, calponin, nonmuscle β- and γ-actins and filamin (Mabuchi *et al* 1997, Small *et al* 1986, Small and Gimona 1997). They establish direct structural and functional contacts between the intracellular cytoskeleton and the extracellular matrix (ECM). The anchoring plaques play an essential role in transmitting forces of contraction-relaxation in the tissue, and act as mechanosensors in gene expression signalling pathways, cell migration, growth and adaptation (Geiger and Ginsberg 1991, Yamada and Geiger 1997, Zamir and Geiger 2001).

Myocytes are arranged into smooth muscle fasciculi, $\simeq 300 \pm 100$ μm, and further are assembled into bundles, $\simeq$ 1–2 mm in length. Electron microscopy and freeze fracture studies have convincingly demonstrated that individual myocytes are interconnected by small and irregular gap junctions. Confocal immunofluorescence, Western blot techniques, transcriptase-PCR reaction and *in situ* hybridisation methods have shown that they are formed mainly by the subunit proteins connexin–43 and 45. These provide the structural basis for cytoplasmic continuity, mediate the movement of ions and small molecules, and support synchronisation and long range integration in the detrusor (Fry *et al* 1999, Wang *et al* 2006, John *et al* 2003, Neuhaus *et al* 2002, Hashitani *et al* 2001, 2004).

Immunohistochemical evidence has demonstrated the presence mainly of collagen types I, III and IV, elastin fibres, laminin, osteopontin, fibronectin, and integrins

(α1–3, $\alpha v\beta 3$, $\alpha 5\beta 1$) in the lamina propria of the normal UB (Wilson *et al* 1996). The three-dimensional hierarchy of folding and coiling of the fine fibrillar matrix in concert with adhesive proteins ensures the syncytial property of the detrusor (Fry *et al* 2004, Rubinstein *et al* 2007). It offers crucial mechanical characteristics such as high compliance, even stress–strain distribution and coordinated phasic contractility during filling and emptying (Nagatomi *et al* 2007, Wognum *et al* 2009). In addition, continuous remodelling of the stromal network allows the organ to respond acutely and efficiently to prolonged periods of strain by adjusting its function and structure through dynamic myocyte-ECM interactions and altering signalling pathways (Aitken and Bägli 2009).

The contractile apparatus of detrusor myocytes consists of thin-actin and thick-myosin filaments, a family of special proteins and kinases, e.g. light chain myosin, tropomyosin, calmodulin, *h*-caldesmon, calponin, myosin light chain kinase and myosin phosphatase. Actin filaments are single helical coils of actin associated with tropomyosin and caldesmon. Myosin filaments are made out of two coil rod-like structure heavy chains with a globular head domain. A principal determinant of the dynamics of contractions is free cytosolic calcium (Ca_i^{2+}) that triggers the cyclic actin–myosin complex formation. Two types of contractions—tonic and phasic—are produced by the detrusor. Thus, during the late stage of bladder filling the muscle generates tonic contractions and undergoes phasic contractions during bladder emptying.

Contractility of the detrusor is controlled by spontaneous and/or induced electrical processes. Their repertoire depends on the balanced function of plasmalemmal ion channels: L- and T-type Ca^{2+}, Ca^{2+}-activated K^+, voltage-dependent K^+, and Cl^- channels. The presence of L- and T-type Ca^{2+} channels in the human UB has been confirmed by electrophysiological and pharmacological studies (Wegener *et al* 2004, Tomoda *et al* 2005, Kajioka *et al* 2002, Uckert *et al* 2000, Elliott *et al* 1996, Badawi *et al* 2006, Hashitani and Brading 2003, Hollywood *et al* 2003). They are formed of five distinct subunits: α_1, α_2, β, δ and γ. The α_1-subunit contains the channel pore, voltage sensor and drug binding sites, while α_2, β, δ and γ-subunits modulate the channel's permeability. L-type channels possess characteristics of long-lasting, high-voltage-dependent channels and ensure the main influx of extracellular calcium ions, Ca_0^{2+}, during depolarisation.

Three subfamilies of T-type Ca^{2+} channels have been identified in the detrusor which differ in their α-subunits. They are activated at low voltage and remain open for a short period of time (Badawi *et al* 2006, Sui *et al* 2001, 2009, Perez-Reyez 2003). Experimental data suggest that the channels are responsible for the generation of spikes and pacemaker activity, playing a key role in regulating the frequency of phasic contractions (Meng and Cha 2009, Brading 2006, Sui *et al* 2006).

Potassium channels constitute a superfamily of four channels: the large Ca^{2+}-activated K^+ (BK_{Ca}), small conductance (SK_{Ca}), voltage-gated (K_v), and ATP-sensitive (K_{ATP}) potassium channels. The BK_{Ca} channel is made of six transmembrane proteins. The channel's sensitivity to calcium and its activity is regulated by phosphorylation of the pore-forming α-subunit. This offers a mechanism whereby cyclic nucleotides and protein kinase C modulate channel function (Tian *et al* 2008). Two types of K_v channels—delayed rectifying and rapidly inactivating—have been identified. They are formed by a single unit of six transmembrane proteins and the pore–hairpin loop. The channels remain uncoupled at low $[Ca_i^{2+}]$ switching to a calcium sensor mode with a rise in intracellular calcium. Together with SK_{Ca} channels, they determine the resting membrane potential, action potential repolarization, excitability and muscle contractility (Herrera *et al* 2000, 2001, Hashitani and Brading 2003, Layne *et al* 2010, Hristov *et al* 2011).

Although intracellular ATP-gated K^+ channels (K_{ATP}) have been implicated in regulating the resting membrane potential and spontaneous mechanical contraction in cells, their overall contribution to electrical activity is considered to be relatively low. The modulation and function of these channels in normal physiological and diseased conditions has not been fully studied yet.

The distinct role of Cl^- channels is unclear due to the uncertainty of their molecular identity. Calcium-activated chloride currents have been recorded on the isolated detrusor smooth muscle cells. They are evoked by the elevation of Ca_i^{2+} and having distinctive biophysical properties (Hartzell *et al* 2005, Chen 2005, Dutzler 2007). However, until the structure of these channels is resolved, we can only speculate on their specific functional role.

The resting membrane potential, V^r, of human bladder smooth muscle cells ranges between −55 mV and −38 mV. Estimated and direct measurements of the input membrane resistance and capacitance of myocytes have shown $R_m \simeq 125 \pm 49$ $M\Omega{\cdot}cm^2$ and $C_m \simeq 1.0\ \mu F\ cm^{-2}$, respectively (Hashitani and Brading 2003, Sui *et al* 2006, Meng and Cha 2009). There is controversy over the existence of spontaneous slow wave activity with some authors (Brading 2006) claiming that intracellular recording from isolated and intact strips of detrusor do not show low amplitude resting potential oscillations consistent with slow waves. By contrast, the traces of simultaneous recordings of mechanical and intracellular electrical activity in human detrusor smooth muscle obtained by Visser and van Mastrigt (1999, 2000, 2001) convincingly demonstrate spontaneous fluctuations of the resting membrane potential of amplitude ~8–10 mV at a wide range of frequencies: $\nu = 0.33$–25 (Hz). The detrusor muscle produces spontaneous action potentials (APs) or spikes of magnitudes ≃ 34–46.5 (mV) and $\nu \simeq$ Z 0.07–0.28 (Hz) (Visser and van Mastrigt 1999, Hashitani and Brading 2003). They occur as single, clusters or bursts of 3–20 action potentials. Each spike has a relatively constant duration, ~1.3 s, a characteristic slow rising phase of depolarisation, ~0.6 s, followed by a fast after hyperpolarization phase, ~0.7 s. Spontaneous action potentials are resistant to tetrodotoxin (TTX), caffeine, ryanodine, thapsigargin and cyclopiazonic acid, suggesting that extrinsic innervation and intracellular calcium stores do not contribute to their generation. However, spikes are abolished by L-type Ca^{2+} channels blockers, e.g. nifedipine, verapamil, or in calcium free solutions, indicating that they are of intrinsic (intramural) origin.

The conduction velocity, v_φ of action potentials in mammals has been evaluated using the electromyographic mapping technique. The results have shown that the maximum v_φ in the rabbit detrusor is 3 cm s^{-1} (Kinder *et al* 1998), depending on the site and physiological status of the organ. Electrical coupling and passive cable properties of detrusor muscle cells from a pig bladder were studied with the two-electrode method (Hashitani and Brading 2003). Although the results are inconclusive, it may be assumed that action potentials have a preferred direction of propagation along the axis of the muscle cell over a short distance. The spread of excitation in the transverse direction is poor. The general harmony of observations with the anatomical structure and distribution of gap junctions suggests that the detrusor syncytium possesses properties of electrical anisotropy. However, no direct attempts to measure preferential conductivity in the human UB have been carried out.

1.3 'Pacemakers'

There is increasing experimental evidence demonstrating that myofibroblasts—interstitial cells (IC)—modulate spontaneous electrical activity of the bladder. Using methods of transmission electron microscopy, immunostaining and *c-kit* receptor labelling, ICs have been found to be abundantly distributed immediately below the urothelium and between detrusor cells and smooth muscle bundles (Kubota *et al* 2011, van der Aa *et al* 2004, Klemm *et al* 1999, McCloskey and Gurney 2002, Hashitani *et al* 2001, Hashitani 2006, McHale *et al* 2006). According to their location, ICs are divided into three subpopulations: (i) boundary IC—adjacent to the boundary of the bladder, (ii) intramuscular IC—scattered among smooth muscle cells within muscle bundles, and (iii) interbundle IC—distributed in connective tissues. They form close connections with intramural nerves and respond positively to various chemical mediators (McCloskey 2010). Based mainly on morphological similarities with interstitial cells of Cajal found in the gastrointestinal tract where their role as pacemakers is 'established', it has been hypothesised that myofibroblasts in the UB act as pacemakers. This view is supported by evidence that application of imatinib mesylate—a selective *c-kit* antagonist—disrupts spontaneous electrical activity in the organ (Kubota *et al* 2004). On the other hand, experiments on single and groups of smooth muscle cells reveal that they are able to produce spontaneous discharges even without ICs. Moreover, it has been shown that Ca^{2+} transients in ICs occur independently of those of smooth muscles even when synchronous calcium waves sweep across muscle bundles (Hashitani *et al* 2004). Therefore, there is reason to believe that ICs play a role in mediating the propagation of action potentials and not in providing the focus for their generation (Hashitani *et al* 2004).

A comparative analysis of the behaviours of isolated cells and muscle strips from different regions of the UB also suggests that the trigone myocytes may serve as the precursor for spontaneous electromechanical activity (Roosen *et al* 2009). However, the concept is based on speculative assumptions about morpho-functional relationships and has not been fully tested experimentally.

The generation of strong regular electrical discharges is essential for the development of coordinated forceful contractions in the UB. It is most likely achieved through the dense intramural para-sympathetic innervation of the wall and the network of intramural ganglia rather than through the syncytial cable properties guaranteed by existing adherens and gap junctions.

1.4 Mechanosensation

A dense nexus of first-order sensory neurons located in the bladder wall detect and relay information to the spinal cord regarding the distension of the organ, its discomfort and/or pain. The mechanosensor as a morphological entity has not yet been established. However, expressed equivocally throughout the bladder tissues and afferent fibres, purinergic ($P2X_{1-7}$ and $P2Y_{1/2/4}$) and transient receptor potential ($TRPV_{1/2/4}$, $TRPM_{7/8}$, $TRPA_1$) channels have been implicated as major basic units in nociception and mechanosensory transduction, in addition to adrenergic (α_2 and $\beta_{2/4}$), cholinergic (muscarinic, μ_1–μ_5 and nicotinic), protease-activated, neuro-trophin (p75NTR), tropomyosin receptor kinases A and B, $CRF_{1/2}$, pituitary adenylyl cyclase-activating polypeptide (PACAP) type 1, vasoactive intestinal polypeptide (VIP), chemokine ($CXCR_4$, $CX3CR_1$) receptors and acid sensing ion channels ($ASIC_{2a/3}$) (Merrill *et al* 2016, Girard *et al* 2017). Despite being primarily involved in mechanosensation, purinergic receptors and TPR channels also regulate neuronal cell growth, remodelling, axon guidance, and growth-cone signalling (Andersson and 2019).

Functionally, four populations: (i) detrusor, (ii) detrusor-mucosal, (iii) mucosal high-, and (iv) mucosal low-responding mechanosensory neurons, have been identified in the guinea pig UB (Zagorodnyuk *et al* 2007). The first group have the receptive fields present in the detrusor whilst in the second they are found in the outer smooth muscle layer and lamina *propria.* Mechanosensitive endings of the third and fourth groups are allied mainly to the urothelium. The detrusor and detrusor-mucosal mechanoreceptors show a low threshold (~1–2 mm) to passive distension, active contraction and rapid stretch sensitivity. It has been argued that the receptor's excitability is also mediated chemically by *in situ* released transmitters such as adenosine triphosphate (ATP), acetylcholine (ACh), substance P (SP) and possibly nitric oxide (NO) and prostaglandin (PrE_2) (Zagorodnyuk *et al* 2006). In response to changes in tension in the wall the receptors modulate the permeability of transmembrane Na^+, Ca^{2+} and other cation channels, thereby leading to the production of high frequency bursting type dendritic action potentials (APs) of amplitude ~0.3 mV. The mucosal low- and high-responding mechanoreceptors are insensitive to distension and weakly activated only by von Frey hair stroking.

Dendritic APs are transferred to the spinal cord neurons via the afferent thin myelinated (Aδ) and unmyelinated (C) fibres of the pelvic, pudendal and hypogastric nerves. Although the current consensus is that under physiological conditions Aδ—fibres convey information on UB filling whereas C-fibres pass primarily noxious stimuli, the results of unbiased single-cell RNA sequencing studies of DRG neurons suggest the existence of both low-threshold mechano- and nociceptor types in Aδ neurons (Fowler *et al* 2008, Usoskin *et al* 2015). The somata of both fibres are located in the DRG at the level of S2-S4 and T11-L2 spinal segments. The ganglia comprise T-shaped pseudo-unipolar neurons: (i) short proximal processes,

20–100 μm in diameter, extending to the secondary-sensory neurons in the intermediate grey matter of the spinal dorsal horn (SDH), and (ii) distal processes, 10–25 cm long and 1 μm thin, projecting towards the bladder where they terminate in the wall as free nerve endings (Rudomin 2002, Emery and Ernfors 2018, Haberberger *et al* 2019). The afferent fibres are unique in transmitting sensory signals which bypass the cell body leading directly to the secondary-neurons in SDH (Amir and Devor 2003).

Immunohistochemical, *in situ* hybridisation, RT-PCR and RNA sequencing *in vitro* studies have confirmed the diverse presence of voltage-dependent sodium ($Na_V1.7$–1.9), calcium ($Ca_V2.2$–2.3), calcium-activated potassium ($K_{Ca}2.1$, $K_{Ca}3.1$), and non-selective cation ($TRPV_1$, $TRPA_1$) channels in almost all first-order DRG neurons (Han *et al* 2015, Boettger *et al* 2002). In addition, there are a variety of neuropeptides, e.g. calcitonin-gene-related-peptide, ATP, SP, galanin, endothelin-1, neurotransmitters and amino acids, e.g. nitric oxide (NO), glutamate (Glu) and γ-amino butyric acid (GABA) synthesised and co-released at the central and peripheral terminals of sensory neurons (Fernández-Montoya and Avendaño 2018, Gu and MacDermott 1997, Nakatsuka and Gu 2001, Davidson *et al* 2014). There is extensive cross-talk among them. Thus results of patch-clamp recordings from SDH neurons indicate the positive role of ATP in modulating glutamatergic signalling whilst the addition of GABA impedes it (Lee *et al* 2002, Nakatsuka and Gu 2001, Vulchanova *et al* 1997, Du *et al* 2017). NO enhances inhibitory glycinergic input and reduces glutamate release from primary afferent terminals through cGMP-protein kinase G and S-nitrosylation of voltage-activated Ca^{2+} channels (Jin *et al* 2011). These, along with high speed nerve pulse conduction, 70–120 m s^{-1}, sensitivity and specificity to the intensity of various stimuli, e.g. innocuous, noxious and mechanical, guarantee the robust linkage of peripheral afferent fibres to secondary-sensory neurons in laminae I and II of the SDH along with the efficient conveyance of polymodal sensory messages.

1.5 Afferent signalling

To date there has been no comprehensive histological classification of the secondary-sensory neurons. Cytoarchitecturally, these are broadly divided into two large groups with regard to the extension of their axons, i.e. those that project to the brain (tract cells) and those remaining within the spinal cord (non-tract cells or interneurons). Morphologically, these are classified according to cellular geometry and aborization of both axons and dendrites. Thus, cells in lamina I have large (10.8–27.3 μm) fusiform, pyramidal or multipolar bodies and axons extending ventralaterally (VL-projection neurons) or ipsilaterally (non-projection neurons). Their dendritic trees are relatively sparse and typically spread medio-laterally and rostro-caudally. By contrast, the cells in lamina II show greater diversity being usually smaller in size (13.2–22 μm) and more regular in shape when compared to lamina I cells (Grundt and Perl 2002, Schoenen 1982). Depending on the directions of elongation of dendrites, the neurons are categorised as isle, medial-lateral, radial, central or vertical types.

Results of tight-seal whole-cell recordings reveal a resting membrane potential of −44 to −62 (mV) and input resistance, 0.15–1 MΩ. Intriguingly, the lamina I and II neurons within a thickness of the spinal cord segment produce putative monosynaptic excitatory/inhibitory postsynaptic potentials (EPSPs/IPSPs) in response to an electrical discharge of a corresponding segmental DRG. Furthermore, conduction velocities (υ) of afferent fibres initiating these responses differ with respect to the category of neurons at which they terminate. Thus for VL-projection, radial, vertical and central neurons $\upsilon = 0.74–0.9\ \mathrm{m\ s^{-1}}$ suggest that inputs are carried by Aδ-fibres, while for non-projecting, islet and media-lateral types, $\upsilon = 0.28–0.43\ \mathrm{m\ s^{-1}}$ indicate that possible connections are through C-fibres (Grundt and Perl 2002).

Chemical profiling of synapses, i.e. partial or total suppression of inward/outward ion currents in the presence of an AMPA receptor antagonist CNQX (6-cyano-7-nitroquinoxaline-2,3-dione), and a selective decrease in outward current after an application of strychnine and bicucullin—Gly and $GABA_A$ receptor antagonists, respectively—suggests the Glu involvement in excitatory, and GABA in inhibitory, transmission (Nakatsuka and Gu 2001, Bao *et al* 1996, 1998). These notions are supported by direct immunocytochemical identification of the transmitter phenotype and through the use of whole-cell patch-clamp recordings taken from synaptically linked neurons (Yasaka *et al* 2010, Zheng *et al* 2010, Lu and Perl 2005). Immunostaining of the axons of cells using VGABA (GABA transporter), glutamate decarboxylase (the rate limiting enzyme for GABA synthesis—GAD) and $GlyT_2$ (the neuronal glycine transporter 2) antibodies has asserted the existence of purely glycinergic, GABAergic synapses and those containing two amino acids (Brumovsky 2013, Todd 2010, Todd and Sullivan 1990, Polgar *et al* 2013).

Lamina I and II SDH neurons generate somatic APs of average amplitude 65–70 mV with four distinct firing patterns: transient, gap, delayed and reluctant (Graham *et al* 2007, Ruscheweyh and Sandkuhler 2002). Each reflects the presence and activity of specific ion channels. For example, gap, delayed and reluctant responses are commonly recorded from excitatory neurons and are mediated by voltage-gated K^+ channels ($K_V4.2$ subunit) (Hu *et al* 2006). Most inhibitory neurons, however, display a tonic or transient pattern of firing at the respective frequencies of 12 Hz, and 0.3–0.6 Hz.

At the time of writing, no neuro-anatomical tract-tracing data was available on projection neurons that relay mechanosensory signals from the human UB to the thalamus and/or directly to the periventricular nuclei in the brain. It is conceivable that messages are passed in ascending spinothalamic tracts. It is also conceivable, based on the fact of their origin, that the projection cells are glutamatergic in nature (Todd 2010). This is supported by single- and double-labelling immunoelectron microscopy and mRNA encoding studies for vesicular glutamate transporter 2 ($VGluT_2$), the H^+/Glu antiporter expressed only in glutamatergic neurons. Results have allowed the identification and visualisation of central axo-somatic and axo-dendritic glutamate containing boutons in the anterior periventricular neurons and in the spinothalamic terminals on the ventral posterior thalamic nuclei (Kiss *et al* 2006, Todd 2017, Graziano *et al* 2008).

1.6 Neuronal assemblies in the hypothalamus and adenohypophysis

1.6.1 Periventricular nucleus

The thin sheet of neurons lining the wall of the third ventricle is termed the periventricular nucleus of the hypothalamus. The PeVN comprises (of) several anatomical regions: rostral (anterior), intermediate and caudal (posterior). The general consensus is that cells in the anterior and intermediate regions have a neuroendocrine function aiding the production and release of somatostatin (SS), neuropeptide Y (NPY), galanin and other hormones, while the main role of those in the caudal region is the regulation of the sympathetic nervous system. The neurons, about 15 μm in diameter, possess a simple dendritic tree morphology and are classified as multipolar, bipolar, and unipolar. The processes run either dorsoventrally or rostro-caudally with a great degree of overlapping among them, thus offering extensive connectivity with other neighbouring nuclei. Neuro-anatomical microscopy studies have indicated a scarcity of direct and indirect interrelations, 5.5%–8.1%, between periventricular and the growth hormone-releasing hormone (GHRH)—positive neurons in the ArcN. The majority of axons, ~78%, project towards and terminate in the median eminence (ME) (Daikoku *et al* 1998, Ducret *et al* 2010). These have been affirmed electrophysiologically by the recording of mono- and polysynaptic responses from ArcN cells (Lanneau *et al* 2000).

Our current knowledge of the morphoarchitecture and neurochemistry of synapses and phenotypic receptor characterisation and distribution in the PeVN neurons has been taken from: dual-label immunofluorescent, *in situ* hybridisation, northern blot analysis, confocal laser microscopic and pharmacological investigations. However, the reader should be reminded that these methods are limited by the number of proteins or transcripts that can be visualised simultaneously and by a strong bias towards known/available markers. Despite the constraints, the results have shown: (i) the co-expression of SS and mRNA for $VGluT_2$ in the cell bodies and axonal terminals in the rostral neurons (Hrabovszky *et al* 2005), (ii) the presence of GAD-positive vesicles in axo-somatic and axo-dendritic synapses (Kakucska *et al* 1988), and (iii) the expression of GH receptor gene in SS-containing neurons (Burton *et al* 1992, Señarís *et al* 1996, Zheng *et al* 1997, Minami *et al* 1998).

Electrophysiologically, all neurons independent of their cytomorphology produce APs of amplitude 60–65 (mV) with a mean duration at half amplitude 0.44 ± 0.02 ms. According to the frequencies, ν, of firing, four patterns have been observed: irregular, bursting, tonic, or tonic-bursting. For example, neurons with tonic activity discharge spikes at $\nu = 2.6 \pm 0.3$ Hz, whilst both bursting and tonic-bursting types fire at smaller rates: 0.43 ± 0.08 and 0.93 ± 0.15 Hz, respectively (Ducret *et al* 2010). Results of acute *in vitro* direct pharmacological modulation studies with α_1-, i.e. methoxamine and phenylephrine, and/or β-agonists, i.e. isoprotenerol, have demonstrated a dose-dependent increase in the excitability of cells, whereas the addition of selective α_1- and β-adrenoceptor antagonists, i.e. prazosin and propranolol, either suppress or reverse the effect (Inenaga *et al* 1986). Although no attempts have been made so far to clone transmembrane ion channels in periventricular neurons, we can theorise regarding the existence and role of diverse Ca^{2+}, K^{+}, Na^{+}, Ca^{2+}-activated K^{+} and Cl^{-} channels.

1.6.2 Arcuate nucleus-median eminence complex

An aggregation of cells located along the third ventricle in the ventromedial hypothalamus is referred to as the arcuate nucleus. As a result of its close proximity to and unique anatomical relationship with the ME, i.e. the existence of the blood–brain-barrier, the ArcN and ME are treated as the ArcN-ME circumventricular organ.

Currently, a complete census of ArcN cell types is not available. Morphologically, small pear-shaped fusiform (~10–15 μm in diameter) and large polygonal (~15–25 μm in diameter) neurons have been described. Fusiform neurons are found predominantly in the medial and dorsal parts of the nucleus and along the ependymal wall of the third ventricle. These possess only one sparsely arborizing dendrite with the axon originating at the opposite pole of the soma. By contrast, polygonal neurons located in the ventral and lateral portions of the nucleus show 4–5 repeatedly branching stem dendrites (Bodoky and Réthelyi 1997, van den Pol and Cassiy 1982, Baccam *et al* 2007). A small number of GHRH-positive cell bodies have short axi that terminate within the ArcN itself and receive synaptic inputs from the arcuate SS-containing neurons. The majority of axonal efferent fibres, 0.5–0.8 μm thick, ascending laterally and dorsolaterally, end in the periventricular zone whilst those descending caudally terminate in the external layer of the ME and, as such, enter a BBB-free area (Bouret *et al* 2004, Balthazar *et al* 2003, Ibata and Watanabe 1977). These axi release GHRH and SS into the fenestrated capillaries of the hypophyseal portal blood system.

Modern molecular 'classification' based on (the) transcriptosome profiling from acutely dissociated ArcN-ME cells combined with a principal component analysis, dimensionality reduction and density based-clustering identifies six distinct types of neuron populations: (i) neuroendocrine cells containing growth hormone-releasing hormone (GHRH), somatostatin, galanin, etc, (ii) tubero-infundibular dopaminergic neurons, and (iii) four types of centrally projecting neurons providing synaptic links and peptidergic inputs to almost the entire hypothalamus (Campbell *et al* 2017). Results of *in situ* hybridisation, single-cell quantitative RT-PCR and double-labelled immunocytochemistry studies in the ArcN-ME and its GHRH-positive neurons have specifically demonstrated the expression of multiple peptides, amines, and neurotransmitters for: (i) ghrelin (Ghr) (Bennett *et al* 1997, Mano-Otagiri *et al* 2006); (ii) SS (Tannenbaum *et al* 1998, Breder *et al* 1992, Stroh *et al* 2009); (iii) IGF-1 (Werther *et al* 1989); (iv) GABA, mainly in the nerve terminals (Hrabovszky *et al* 2005, Betley *et al* 2013, Campbell *et al* 2017); (v) NPY (Sun and Miller 1999); (vi) ACh (Baccam *et al* 2007, Gautam *et al* 2009, Jeong *et al* 2016), and (vii) co-expression of: ACh and GABA; ACh and Glu; dopamine and GABA (Baccam *et al* 2007, Jeong *et al* 2016, Zhang and van den Pol 2015).

Such co-localisation of multiple neurotransmitters, peptides and hormones indicate the key role the ArcN-ME complex plays in integrating peripheral blood-borne endocrine and metabolic signals, whilst other sensory inputs provided through the multilevel synaptic circuitry, relay further processed information to other structures in the brain.

Membrane properties of ArcN-ME neurons examined in the current-clamp mode have revealed a mean resting potential of −51 to −53 mV with an input resistance of 1.7–2.0 GΩ. The amplitude of APs triggered by a transient injection of current are in the range of 59–63.5 mV with the time to peak and half width measuring 2.4–3 ms and 1.2–1.4 ms, respectively. These show the bursting firing at a wide range of frequencies from 0.83 to 10 (Hz) (Balthazar *et al* 2003). Surprisingly little is known about the transmembrane ion channels *per se* and their electrophysiological characteristics. At this stage we can only deduce, from indirect assessments, the presence of a population of: (i) K^+—it has been shown that G-protein-gated inwardly rectifying potassium currents are induced by SS (Yang *et al* 2012. Osterstock *et al* 2016); (ii) Ca^{2+}—*in vitro* studies have revealed that SS reduces neuronal Ca^{2+} APs and Ca^{2+} currents by inhibiting N and P/Q type calcium channels and activating NPY receptors (Weckbecker *et al* 2003) whilst Ghr and Glu excites voltage-dependent Ca^{2+} channels (Sun and Miller 1999, Osterstock *et al* 2010, Niciu *et al* 2012); (iii) Ca^{2+}-activated K^+ (Lahlou *et al* 2004); (iv) Cl^-—the treatment with GABA induces chloride efflux (Zemkova *et al* 2008), and other common ion channels.

1.6.3 Adenohypophysis (somatotrpohs)

Somatotrophs are a group of acidophilic cells located in the lateral wings of the anterior pituitary gland (adenohypophysis or pars anterior) that synthesise and secret growth hormone. Their mophological and morphometric evaluation reveal rounded polygonal, triangular or oval shaped cells with a surface area of 55.6 ± 2.9 μm^2 (porcine), 103.7 ± 4.1 μm^2 (rat) and a volume of ~8086 ± 249 μm^3 (Engström and Sävendahl 1995, Jiménez-Reina *et al* 2000). The roughly round or oval nuclei and intracellular organelles are positioned centrally or slightly eccentrically with GH-positive large dense-core vesicles/granules (LDCVs) evenly distributed along the periphery close to the cellular membrane. Based on their number and size, three types of somatoptrophs are distinguished: type I contain large secretory granules ~350 nm in diameter, type II are filled with mixed size granules of 150–350 nm, whilst type III comprise small granules with a diameter measuring ~150 nm (Kurosumi *et al* 1986). Ultrastructure, functional heterogeneity and cell size are all influenced by physiological state, gender and age.

The adenohypophysis is unique in its means of communication and exchange of endocrine signalling with the ArcN-ME complex. These are enabled via the hypophyseal portal blood system—the fine network of contiguous fenestrated microvessels. Arranged in sequence as the primary and secondary plexi, these provide the arterial blood supply to the ArcN-ME complex and further down to the pars anterior. Such a richly intertwined capillary arrangement facilitates both the rapid and efficient transport of hormones, i.e. GHRH and SS among others, to the adenohypophyseal cells and in particular to the somatotrophs. Interestingly, both GHRH mRNA and newly synthesised GHRH are found in somatotrophs although the authors do not provide any compelling proof to support the fact stated (Freeman *et al* 2008).

Somatotrophs have a resting membrane potential of −62 to −65 mV. The cells are excitable and produce spontaneous and induced action potentials. The pattern of spontaneous APs is stereotypical and occurs at a frequency of 0.2 Hz. Instantly depolarising the membrane by 20–25 mV, they generate bursts of low amplitude spikes, 15–20 mV, on the crests of the plateau. The treatment of somatotrophs with GHRH causes regular high frequency discharges of APs of amplitude 40–45 mV observed over the action period of the hormone (Fletcher *et al* 2016). Such specific electrical activity is defined by the dynamic context of the common core of T- and L-type Ca^{2+}; the fast-delayed rectifier (K_{dr}), inwardly rectifying (K_{ir}), slow non-inactivating and transient voltage-gated K^{+}; the voltage-activated Na^{+}; and the leak Cl^{-} transmembrane ion channels (Yang and Chen 2007, Fletcher *et al* 2016, Stojilkovic *et al* 2010).

The last anatomical element in the system under consideration is the hepatocytes. The majority of IGF-1 and the associated insulin-like growth factor binding proteins ($IGFBP_{1-6}$) are continuously produced and secreted in the liver (75% of the total amount) in response to GH stimulation whilst the rest (25%) is made in extrahepatic tissues (Ohisson *et al* 2009).

References

Aitken K J and Bägli D J 2009 The bladder extracellular matrix. Part I: Architecture, development and disease *Nat. Rev. Urol.* **6** 596–611

Amir R and Devor M 2003 Electrical excitability of the soma of the sensory neurons is required for spike invasion of the soma, but not for through-conduction *Biophys. J.* **84** 2181–91

Andersson K A 2019 T P R channels as lower urinary tract sensory targets *Med. Sci.* **7** 67–78

Ansari M S, Gulia A, Srivastava A and Kapoor R 2010 Risk factors for progression to end-stage renal disease in children with posterior urethral valves *J. Pediatr. Urol.* **6** 261–4

Apodaca G, Balestreire E and Birder L A 2007 The uroepithelial-associated sensory web *Kidney Int.* **72** 1057–64

Baccam N, Alonso G, Costecalde T, Fontanaud P, Molino F, Robinson I C A F, Mollard P and Mery P-F 2007 Dual-level afferent control of growth hormone-releasing hormone (GHRH) neurons in GHRH-green fluorescent protein transgenic mice *J. Neurosci.* **27** 1631–41

Badawi J K, Li H, Langbein S, Kwon S T, Kamp S and Bross S 2006 Inhibitory effects of L- and T-type calcium antagonists on contractions of human detrusor muscle *Eur. J. Clin. Pharmac.* **62** 347–54

Balthazar N, Mery P-F, Magoulas C B, Mathers K E, Martin A, Mollard P and Robinson I C A F 2003 Growth hormone-releasing hormone (GHRH) neurons in GHRH-enhanced green fluorescent protein transgenic mice: a ventral hypothalamic network *Endocrine* **144** 2728–40

Bao J, Li J and Perl E R 1996 Primary afferent inhibition of spinal laminae I and II neurons *Soc. Neurosci. Abst.* **22** 878

Bao J, Li J and Perl E R 1998 Differences in Ca^{2+} channels governing generation of miniature and evoked excitatory synaptic currents in spinal laminae I and II *J. Neurosci.* **18** 8740–50

Bennett P A, Thomas G B, Howard A D, Feighner S D, van der Ploeg L H, Smith R G and Robinson I C 1997 Hypothalamic growth hormone secretagogue-receptor (GHS-R) expression is regulated by growth hormone in the rat *Endocrine* **138** 4552–7

Betley J N, Cao Z F, Ritola K D and Sternson S M 2013 Parallel, redundant circuit organization for homeostatic control of feeding behaviour *Cell* **155** 1337–50

Bodoky M and Réthelyi M 1997 Dendritic arborization and axon trajectory of neurons in the hypothalamic arcuate nucleus of the rat *Exp. Brain. Res.* **28** 543–55

Boettger M K *et al* 2002 Calcium-activated potassium channel SK1- and IK1-like immunoreactivity in injured human sensory neurons and its regulation by neurotrophic factors *Brain* **125** 252–63

Bouret S G, Draper S J and Simerley R B 2004 Formation of projection pathways from the arcuate nucleus of the hypothalamus to hypothalamic regions implicated in the neural control of feeding behavior of mice *J. Neurosci.* **24** 2797–805

Brading A F 2006 Spontaenous activity of lower urinary tract smooth muscles: correlation between ion channels and tissue function *J. Physiol.* **570** 13–22

Breder C D, Yamada Y, Yasuda K, Seino S, Saper C B and Bell G 1992 Differential expression of somatostatin receptor subtypes in brain *J. Neurosci.* **12** 3920–34

Brumovsky P R 2013 VGLUTs in peripheral neurons and the spinal cord: time for a review *ISRN Neurol* **2013** 829753

Burton K A, Kabinting E B, Clifton D K and Steiner R A 1992 Growth hormone receptor messenger ribonucleic acid distribution in the adult rat brain and its colocalization in hypothalamic somatostatin neurons *Endocrine* **131** 958–63

Campbell J N *et al* 2017 A molecular census of arcuate hypothalamus and median eminence cell types *Nat. Neurosci.* **20** 484–96

Campbell-Walsh 2007 *Urology* 9th edn (Amsterdam: Elsevier)

Chen T-Y 2005 Structure and function of CLC channels *Ann. Rev. Physiol.* **67** 809–39

Daikoku S, Hisano S, Kawano H, Chikamori-Aoyama M, Katgotani Y, Zhang R and Chihara K 1998 Ultrastructural evidence of neuronal regulation on growth hormone secretion *Neuroendocrine* **47** 405–15

Davidson S, Copits B A, Zhang J, Page G, Ghetti A and Gereau R W 2014 Human sensory neurons: Membrane properties and sensitization by inflammatory mediators *Pain* **155** 1861–70

DeLancey J, Gosling J A, Creed K E, Dixon J, Delmas V, Landon D and Norton P 2002 Gross anatomy and cell biology of the lower urinary tract *Incontinence, Sec Int Cons Incontinence* ed P Abrams, L Cardozo, KhouryS and A Wein (Plymouth, MA: Health Publ)

Du X *et al* 2017 Local GABAergic signaling within sensory ganglia controls peripheral nociceptive transmission *J. Clin. Invest.* **127** 1741–56

Ducret E, Gaidamaka G and Herbison A E 2010 Electrical and morphological characteristics of anteroventral periventricular nucleus kisspeptin and other neurons in the female mouse *Endocrine* **151** 2223–32

Dutzler R 2007 A structural perspective on Cl channel and transporter function *FEBS Lett.* **581** 2839–44

Elbadawi A 1991 Anatomy and innervation of the vesicourethral mucular unit of micturition *Clinical Neurourology* (London: Little Brown) pp 5–23

Elliott R A, Norman R I, Parker S G, Whitaker R P and Castleden C M 1996 Effect of treatment with calcium antagonists *in vitro* and *in vivo* on the contractile response of isolated rat and human detrusor muscle *Clin. Sci.* **91** 467–74

Emery E C and Ernfors P 2018 Dorsal root ganglion neuron types and their functional specialization *The Oxford Handbook of the Neurobiology of Pain* ed J N Wood (Oxford: Oxford University Press)

Engström K G and Sävendahl L 1995 Cell volume and shape oscillations in rat tape-II somatotrophs at hypotonic conditions *Catometry* **20** 7–13

Fernández-Montoya J and Avendaño N P 2018 The glutamatergic system in primary somatosensory neurons and its involvement in sensory input-dependent plasticity *Int. J. Mol. Sci.* **19** 69

Fletcher P, Bertram R and Tabak J 2016 From global to local: exploring the relationship between parameters and behaviors in models of electrical excitability *J. Comput. Neurosci.* **40** 331–45

Fowler C J, Griffiths D and de Groat W C 2008 The neural control of micturition *Nat. Rev. Neurosci.* **9** 453–66

Freeman M E, David R, Grattan D R and Houpt T A 2008 The hypothalamus *Neuroscience in Medicine* 3rd edn, ed P M Conn (Totowa, NJ: Humana Press) pp 301–58

Fry C H, Cooklin M, Birns J and Mundy A R 1999 Measurement of intracellular electrical coupling in guinea-pig detrusor smooth muscle *J. Urol.* **161** 660–4

Fry C H, Hussain M, McCarthy C, Ikeda Y, Sui G P and Wu C 2004 Recent advances in detrusor muscle function *Scand. J. Urol. Nephrol. Suppl.* **215** 20–5

Gautam G *et al* 2009 Neural M_3 muscarinic acetylcholine receptors are essential for somatotroph proliferation and normal somatic growth *Proc. Natl Acad. Sci. USA* **106** 6398–403

Geiger B and Ginsberg D 1991 The cytoplasmic domain of adherenstype junctions *Cell Motil. Cytoskel.* **20** 1–6

Girard B M, Tooke K and Vizzard M A 2017 PACAP/receptor system in urinary bladder dysfunction and pelvic pain following urinary bladder inflammation or stress *Front. Syst. Neurosci.* **11** 1–23

Graham B A, Brichta A M and Callister R J 2007 Moving from an averaged to specific view of spinal cord pain processing circuits *J. Neurophysiol.* **98** 1057–63

Graziano A, Liu X-B, Murray K D and Jones E G 2008 Vesicular glutamate transporters define two sets of glutamatergic afferents to the somatosensory thalamus and two thalamocortical projections in the mouse *J. Comp. Neurol.* **507** 1258–76

Grundt T J and Perl E R 2002 Correlations between neuronal morphology and electrophysiological features in the rodent superficial dorsal horn *J. Physiol.* **540** 189–207

Gu J G and MacDermott A B 1997 Activation of ATP P2X receptors elicits glutamate release from sensory neuron synapses *Nature* **389** 749–53

Haberberger R V, Barry C, Dominguez N and Matusica D 2019 Human dorsal root ganglia *Front. Cell Neurosci.* **13** 1–17

Hakenberg O W, Linne C, Manseck A and Wirth M P 2000 Bladder wall thickness in normal adults and men with mild lower urinary tract symptoms and benign prostatic enlargement *Neurourol. Urodyn.* **19** 585–93

Han C, Estacion M, Huang J, Vasylyev D, Zhao P, Dib-Hajj S D and Waxman S G 2015 Human Na(v)1.8: enhanced persistent and ramp currents contribute to distinct firing properties of human DRG neurons *J. Neurophysiol.* **113** 3172–85

Hartzell C, Putzier I and Arreola J 2005 Calcium-activated chloride channels *Ann. Rev. Physiol.* **67** 719–58

Hashitani H 2006 Interaction between interstitial cells and smooth muscles in the lower urinary tract and penis *J. Physiol.* **576** 707–14

Hashitani H and Brading A F 2003 Electrical properties of detrusor smooth muscles from the pig and human urinary bladder *Br. J. Pharmac.* **140** 146–58

Hashitani H and Brading A F 2003 Ionic basis for the regulation of spontaneous excitation in detrusor smooth muscle cells of the guinea-pig urinary bladder *Br. J Pharmac.* **140** 159–69

Hashitani H, Fukuta H, Takano H, Klemm M and Suzuki H 2001 Origin and propagation of spontaneous excitation in smooth muscle of the guinea-pig urinary bladder *J. Physiol.* **530** 273–86

Hashitani H, Yanai Y and Suzuki H 2004 Role of interstitial cells and gap junctions in the transmission of spontaneous Ca^{2+} signals in detrusor smooth muscles of the guinea-pig urinary bladder *J. Physiol.* **559** 567–81

Herrera G M, Heppner T J and Nelson M T 2000 Regulation of urinary bladder smooth muscle contractions by ryanodine receptors and BK and SK channels *Am J Physiol Regul Integr. Comp. Physiol.* **279** R60–68

Herrera G M, Heppner T J and Nelson M T 2001 Voltage dependence of the coupling of Ca^{2+} sparks to BK_{Ca} channels in urinary bladder smooth muscle *Am. J. Physiol. Cell Physiol.* **280** C481–90

Hollywood A M, Woolsey S, Walsh I K, Keane P F, McHale N G and Thornbury K D 2003 T- and L-type Ca^{2+} currents in freshly dispersed smooth muscle cells from the human proximal urethra *J. Physiol.* **550** 753–64

Hrabovszky E, Turi G F and Liposits Z 2005 Presence of vesicular glutamate transporter-2 in hypophysiotropic somatostatin but not growth hormone-releasing hormone neurons of the male rat *Eur. J. Neurosci.* **21** 2120–6

Hristov K L, Chen M, Kellett W F, Rovner E S and Petkov G V 2011 Large-conductance voltage- and Ca^{2+}- activated K^{+}- channels regulate human detrusor smooth muscle function *Am. J. Physiol. Cell Physiol.* **301** C903–12

Hu H J, Carrasquillo Y, Karim F, Jung W E, Nerbonne J M, Schwarz T L and Gereau R W 4th 2006 The Kv4.2 potassium channel subunit is required for pain plasticity *Neuron.* **50** 89–100

Ibata Y and Watanabe K 1977 A morphological survey of the median eminence: fluorescence histochemistry, electron microscopy and immunohistochemistry *Arch. Hist. Jap.* **40** 303–15

Inenaga K, Dyball R E J, Okuya S and Yashamita H 1986 Characterization of hypothalamic noradrenaline receptors in the supraoptic nucleus and periventricular region on the paraventricular nucleus of mice *in vitro Brain Res.* **369** 37–47

Jeong J H, Woo Y J, Chua S and Jo Y H 2016 Single-cell gene expression analysis of cholinergic neurons in the arcuate nucleus of the hypothalamus *PLoS One* **11** e0162839

Jiménez-Reina L, Cañete R, Cepeda T and Bernal G 2000 Morphological changes to somatotroph cells and *in vivo* individual GH release, in male rats treated with recombinant human GH *Histol. Histopathol.* **15** 375–83

Jin X-G, Chen S-R, Co X-H, Li L and Pan H-L 2011 Nitric oxide inhibits nociceptive transmission by differentially regulating glutamate and glycine (Gly) release to spinal dorsal horn neurons *J. Biol. Chem.* **286** 33190–202

John H, Wang X, Wehrli E, hauri D and Maake C 2003 Evidence of gap junctions in the stable nonobstructed human bladder *J. Urol.* **169** 745–9

Kajioka S, Nakayama S, McMurray G, Abe K and Brading A F 2002 Ca^{2+} channel properties in smooth muscle cells of the urinary bladder from pig and human *Eur. J. Pharmac.* **443** 19–29

Kakucska I, Tappaz M L, Gaal G, Stoekel M E and Makara G B 1988 Gabaergic innervation of somatostatin-containing neurosecretory cells of the anterior periventricular hypothalamic area: a light and electron microscopy double immunolabeling study *Neuroscience* **25** 585–93

Kinder M V, Gommer E D, Janknegt R A and van Waalwijk van Doorn E S C 1998 A method for the electromyographic mapping of the detrusor smooth muscle *Arch. Physiol. Biochem.* **105** 673–90

Kiss J, Csaba Z, Csáki A and Halász B 2006 Glutamatergic innervation of growth hormone-releasing-containing neurons in the hypothalamic arcuate nucleus and somatostatin-containing neurons in the anterior periventricular nucleus of the rat *Brain Res. Bull.* **70** 278–88

Klemm M F, Exintaris B and Lang R J 1999 Identification of the cells underlying pacemaker activity in the guinea-pig upper urinary tract *J. Physiol.* **519** 867–84

Kubota Y, Kajioka S, Biers S M, Yokota E, Kohri K and Brading A F 2004 Investigation of the effect of the c-kit inhibitor Glivec on isolated guinea-pig detrusor preparations *Auton. Neurosci.* **115** 64–73

Kubota Y, Kojima Y, Shibata Y, Imura M, Sasaki S and Kohri K 2011 Role of kit-positive interstitial cells of Cajal in the urinary bladder and possible therapeutic target for overactive bladder *Adv. Urol.* **2011** 816342

Kurosumi K, Koyama T and Tosaka H 1986 Three types of growth hormone cells of the rat anterior pituitary as revealed by immunoelectron microscopy using a colloidal gold-antibody method *Arch. Hist. Jap.* **49** 227–42

Lahlou H, Guilleermet J, Hortala M, Vernejoul F, Pyronnet S, Bousquet C and Susini C 2004 Molecular signalling of somatostatin receptors *Ann. New York Acad. Sci.* **1014** 121–31

Lanneau C, Peineau S, Petit F, Epelbaum J and Gardette R 2000 Somatostatin modulation of excitatory transmission between periventricular and arcuate hypothalamic nuclei *in vitro Am. J. Phys.* **84** 1464–74

Layne J J, Nausch B, Olesen S-P and Nelson M T 2010 BK channel activation by NS11021 decreases excitability and contractility of urinary bladder smooth muscle *Am. J. Physiol. Regul. Integr. Comp. Physiol.* **298** R378–84

Lee C J, Bardoni R, Tong C-K, Engelman H S, Joseph D J, Magherini P C and MacDermott A B 2002 Functional expression of AMPA receptors on central terminals of rat dorsal root ganglion neurons and presynaptic inhibition of glutamate release *Neuron.* **35** 135–46

Lopez Pereira P, Martinez Urrutia M J, Espinosa L, Lobato R, Navarro M and Jaureguizar E 2002 Bladder dysfunction as a prognostic factor in patients with posterior urethral valves *BJU Int.* **90** 308–11

Lu Y and Perl E R 2005 Modular organization of excitatory circuits between neurons of the spinal superficial dorsal horn (laminae I and II) *J. Neurosci.* **25** 3900–7

Mabuchi K, Li B, Ip W and Tao T 1997 Association of calponin with desmin intermediate filaments *J. Biol. Chem.* **272** 22662–6

Malmqvist U, Arner A and Uvelius B 1991a Contractile and cytoskeletal proteins in smooth muscle during hypertrophy and its reversal *Am. J. Physiol. Cell. Physiol.* **260** C1085–93

Malmqvist U, Arner A and Uvelius B 1991b Cytoskeletal and contractile proteins in detrusor smooth muscle from bladders with outlet obstruction—a comparative study in rat and man *Scand. J. Urol. Nephrol.* **25** 261–7

Mano-Otagiri A, Nemoto T, Sekino A, Yamauchi N, Shuto Y, Sugihara H, Oikawa S and Shibasaki T 2006 Growth hormone-releasing hormone (GHRH) neurons in the arcuate nucleus (Arc) of the hypothalamus are decreased in transgenic rats whose expression of ghrelin receptor is attenuated: evidence that ghrelin receptor is involved in the up-regulation of GHRH expression in the *Arc Endocr.* **147** 4093–103

Martin A F, Bhatti S, Pyne-Geithman G J, Farjah M, Manaves V, Walker L, Franks R, Strauch A R and Paul R J 2007 Expression and function of COOH terminal myosin heavy chain isoforms in mouse smooth muscle *Am. J. Physiol. Cell Physiol.* **293** C238–45

McCloskey K D 2010 Interstitial cells in the urinary bladder—localization and function *Neurourol. Urodyn.* **29** 82–7

McCloskey K D and Gurney A M 2002 Kit positive cells in the guinea pig bladder *J. Urol.* **168** 832–6

McHale N G, Hollywood M A, Sergeant G P, Shafei M, Thornbury K T and Ward S M 2006 Organization and function of ICC in the urinary tract *J. Physiol.* **576** 689–94

Meng E and Cha T-L 2009 Spontaneous activity of the detrusor and its involvement in generating bladder overactivity *JTUA* **20** 57–62

Merrill L, Gonzalez E J, Girard B M and Vizzard M A 2016 Receptors, channels, and signalling in the urothelial sensory system in the bladder *Nat. Rev. Urol.* **13** 193–204

Minami S, Kamegai J, Sugihara H, Suzuki N and Wakabayashi I 1998 Growth hormone inhibits its own secretion by acting on the hypothalamus through its receptors on neuropeptide Y neurons in the arcuate nucleus and somatostatin neurons in the periventricular nucleus *Endoc. J.* **45** S19–26

Murakumo M, Ushiki T, Abe K, Matsumura K, Shinno Y and Koyanagi T 1995 Three-dimensional arrangement of collagen and elastin fibers in the human urinary bladder: A scanning electron microscopic study *J. Urol.* **154** 251–6

Nagatomi J, Toosi K K, Chancellor M B and Sacks M S 2007 Contribution of the extracellular matrix to the viscoelastic behavior of the urinary bladder wall *Biomech. Model. Mechanobiol.* **7** 395–404

Nakatsuka T and Gu J G 2001 ATP P2X Receptor-mediated enhancement of glutamate release and evoked EPSCs in dorsal horn neurons of the rat spinal cord *J. Neurosci.* **21** 6522–31

Neuhaus J, Weimann A, Stolzenburg J-U, Wolburg H, Horn L-C and Dorscher W 2002 Smooth muscle cells from human urinary bladder express connexin 43 *in vivo* and *in vitro Worl. J. Urol.* **20** 250–4

Niciu M J, Kelmendi B and Sanacora G 2012 Overview of glutamatergic neurotransmission in the nervous system *Pharmacol. Biochem. Behav.* **100** 656–64

Osterstock G *et al* 2010 Ghrelin stimulation of growth hormone-releasing hormone neurons is direct in the arcuate nucleus *PLoS One* **5** e9159

Osterstock G *et al* 2016 Somatostatin triggers rhythmic electrical firing in hypothalamic GHRH neurons *Sci. Rep.* **6** 24394

Perez-Reyez E 2003 Molecular physiology of low-voltage-activated T-type calcium channels *Physiol. Rev.* **83** 117–61

Polgar E, Durrieux C, Hughes D I and Todd A J 2013 A quantitative study of inhibitory interneurons in laminae I-III of the mouse spinal dorsal horn *PLoS One* **8** e78309

Roosen A, Wu C, Sui G, Chowdhury B A, Patel P M and Fry C H 2009 Characteristics of spontaneous activity in the bladder trigone *Eur. J. Urol.* 346–54

Rubinstein M, Sampaio F J and Costa W S 2007 Stereological study of collagen and elastic system in the detrusor muscle of bladders from controls and patients with intravesicular obstruction *Int. Braz. J. Urol.* **33** 33–339

Rudomin P 2002 Selectivity of the central control of sensory information in the mammalian spinal cord *Adv. Exp. Med. Biol.* **508** 157–70

Ruscheweyh R and Sandkuhler J 2002 Lamina-specific membrane and discharge properties of rat spinal dorsal horn neurones *in vitro J. Physiol.* **541** 231–44

Schoenen J 1982 The dendritic organization of the human spinal cord: The dorsal horn *Neuroscience* **7** 2057–87

Señarís R M, Lago F, Coya R, Pineda J and Diéguez C 1996 Regulation of hypothalamic somatostatin, growth hormone-releasing hormone, and growth hormone receptor messenger ribonucleic acid by glucocorticoids *Endocrine* **137** 5236–41

Small J V, Furst D O and De Mey J 1986 Localization of filamin in smooth muscle *J. Cell Biol.* **102** 210–20

Small J V and Gimona M 1997 The cytoskeleton of the vertebrate smooth muscle cell *Acta Physiol. Scand.* **164** 341–8

Stojilkovic S S, Tabak J and Bertram R 2010 Ion channels and signaling in the pituitary gland *Endocr. Rev.* **31** 845–915

Stroh T, van Schouwenburg M R, Beaudet A and Tannenbaum G S 2009 Subcellular dynamics of somatostatin receptor subtype 1 in the rat arcuate nucleus: receptor localization and synaptic connectivity vary in parallel with the ultradian rhythm of growth hormone secretion *J. Neurosci.* **29** 8198–205

Sui G, Fry C H, Malone-Lee J and Wu C 2009 Aberrant Ca^{2+} oscillations in smooth muscle cells from overactive human bladders *Cell Calcium* **45** 456–64

Sui G P, Wu C and Fry C H 2001 Inward Ca^{2+} currents in cultured and freshly isolated detrusor muscle cells—evidence of a T-type Ca^{2+} current *J. Urol.* **165** 627–31

Sui G P, Wu C, Severs N, Newgreen D and Fry C H 2006 The association between T-type Ca^{2+} current and outward current in isolated human detrusor cells from stable and overactive bladders *BJU Internat.* **99** 436–41

Sun L and Miller R J 1999 Multiple neuropeptide Y receptors regulate K^+ and Ca^{2+} channels in acutely isolated neurons from the rat arcuate nucleus *J. Neurophysiol.* **81** 1391–403

Tannenbaum G S, Zhang W H, Lapointe M, Zeitler P and Beaduet A 1998 Growth hormone-releasing hormone neurons in the arcuate nucleus express both sst1 and sst2 somatostatin receptor genes *Endocrine* **139** 1450–3

Tian L, McClafferty H, Chen L and Shipston M J 2008 Reversible tyrosine protein phosphorylation regulates large conductance voltage- and calcium-activated potassium channels via cortactin *J. Biol. Chem.* **283** 3076–86

Todd A J 2010 Neuronal circuitry for pain processing in the dorsal horn *Nat. Rev. Neurosci.* **11** 823–36

Todd A J 2017 Identifying functional populations among the interneurons in laminae I-III of the spinal dorsal horn *Mol. Pain* **113** 1–19

Todd A J and Sullivan A C 1990 Light microscope study of the coexistence of GABA-like and glycine-like immunoreactivities in the spinal cord of the rat *J. Comp. Neurol.* **296** 496–505

Tomoda T, Aishima M, Takano N, Nakano T, Seki N, Yonemitsu Y, Sueishi K, Naito S, Ito Y and Teramoto N 2005 The effects of flavoxate hydrochloride on voltage-dependent L-type Ca^{2+} currents in human urinary bladder *Br. J. Pharmac.* **146** 25–32

Uckert S, Stief C G, Odenthal K P, Truss M C, Lietz B and Jonas U 2000 Responses of isolated normal human detrusor muscle to various spasmolytic drugs commonly used in the treatment of the overactive bladder *Arzneimittelforschung* **50** 456–60

Usoskin D *et al* 2015 Unbiased classification of sensory neuron types by large single-cell RNA sequencing *Nat. Neurosci.* **18** 145–53 van den Pol A N and Cassiy J R 1982 The hypothalamic arcuate nucleus of rat-A quantitative golgi analysis *J. Comp. Neurol.* **204** 65–98

van der Aa F, Roskams T, Byweet W, Ost D, Bogaert G and De Ridder D 2004 Identification of kit positive cells in the human urinary bladder *J. Urol.* **171** 2492–6

Visser A J and van Mastrigt R 1999 Intracellular recording of spontaneous electrical activity in human urinary bladder smooth muscle strips *Arch. Physiol. Biochem.* **107** 257–70

Visser A J and van Mastrigt R 2000 The role of intracellular and extracellular calcium in mechanical and intracellular electrical activity of human urinary bladder smooth muscle *Urol. Res.* **28** 260–8

Visser A J and van Mastrigt R 2001 Intracellular electrical activity in human urinary bladder smooth muscle: the effect of high sucrose medium *Urol. Int.* **66** 205–11

Vulchanova L, Riedl M S, Shuster S J, Buell G, Surprenant A, North R A and Elde R 1997 Immunohistochemical study of the $P2X_2$ and $P2X_3$ receptor subunits in rat and monkey sensory neurons and their central terminals *Neuropharmacology* **36** 1229–42

Wang H Z, Brink P R and Christ G J 2006 Gap junction channel activity in short-term cultured human detrusor myosyte cell pairs: gating and unitary conductances *Am. J. Physiol. Cell Physiol.* **291** C1366–76

Weckbecker G, Lewis I, Albert R, Schmidt H A, Hoyer D and Bruns C 2003 Opportunities in somatostatin research: biological, chemical and therapeutic aspects *Nat. Rev. Drug Disc* **2** 999–1017

Wegener J W *et al* 2004 An essential role of Cav1.2 L-type calcium channel for urinary bladder function *FASEB J.* **18** 1159–61

Werther G A, Hogg A, Oldfield B J, McKinley M J, Figdor R and Mendelsohn F A O 1989 Localization and characterization of insulin-like growth factor-1 receptors in rat brain and pituitary gland using *in vitro* autoradiography and computerized densitometry* a distinct distribution from insulin receptors *J. Neuroendocr.* **1** 369–77

Wilson C B, Leopard J, Cheresh D A and Nakamura R M 1996 Extracellular matrix and integrin composition of the normal bladder wall *World J. Urol.* **14** S30–7

Wognum S, Schmidt D E and Sacks M S 2009 On the mechanical role of de novo synthesized elastin in the urinary bladder wall *Biomech. Eng.* **131** 101018

Yamada K M and Geiger B 1997 Molecular interactions in cell adhesion complexes *Curr. Opin. Cell Biol.* **9** 76–85

Yang R, Amir J, Liu H and Chaqour B 2008 Mechanical strain activates a program of genes functionally involved in paracrine signaling of angiogenesis *Physiol. Genomics* **36** 1–14

Yang S K and Chen C 2007 Involvement of somatostatin receptor subtypes in membrane ion channel modification by somatostatin in pituitary somatotropes *Clin. Exp. Pharmac. Physiol.* **34** 1221–7

Yasaka T, Tiong S Y X, Hughes D I, Riddell J S and Todd A J 2010 Populations of inhibitory and excitatory interneurons in lamina II of the adult rat spinal dorsal horn revealed by a combined electrophysiological and anatomical approach *Pain* **151** 475–88

Zagorodnyuk V P, Costa M and Brookes S J 2006 Major classes of sensory neurons to the urinary bladder *Auton. Neurosci.* **126–127** 390–7

Zagorodnyuk V B, Gibbins I L, Costa M, Brookes S J H and Gregory S J 2007 Properties of the major classes of mechanoreceptors in the guinea pig bladder *J. Phys.* **585** 147–63

Zamir E and Geiger B 2001 Molecular complexity and dynamics of cell-matrix adhesions *J. Cell Sci.* **114** 3583–90

Zemkova H W, Bjelobaba I, Tomic M, Zemkova H and Stojilkovic S S 2008 Molecular, pharmacological and functional properties of GABA(A) receptors in anterior pituitary gland *J. Physiol.* **586** 3097–111

Zhang X and van den Pol 2015 Dopamine/tyrosine hydroxylase neurons of the hypothalamic arcuate nucleus release GABA, communicate with dopaminergic and other arcuate neurons, and respond to dynorphin, met-enkephalin, and oxytocin *J. Neurosci.* **35** 14966–82

Zheng H, Bailey A, Jiang M-H, Honda K, Chen H Y, Trumbauer M E, van der Ploeg L H T, Schaeffer J S, Leng G and Smith R G 1997 Somatostatin receptor subtype 2 knockout mice are refractory to growth hormone-negative feedback on arcuate neurons *Mol. Endoc.* **11** 1709–17

Zheng J, Lu Y and Perl E R 2010 Inhibitory neurons of the spinal substantia gelatinosa mediate interaction of signals from primary afferents *J. Physiol.* **588** 2065–75

Chapter 2

Continual models of the bladder tissue

A new broom sweeps clean but an old broom knows the corners

— Virgin Islander

An overview of the urinary bladder wall within the general framework of soft tissue mechanics is provided. Histoarchitectural correlations are drawn with the dynamics of stress–strain development in the tissue under simple and complex loading. Mathematical models of the tissue as a single and multiphase continuum are presented and discussed.

2.1 Single-phase models

Over the years, surprisingly little attention has been given to the problem of constructing constitutive models of the urinary bladder (UB) wall. The most common type of mathematical models used for soft tissues are phenomenological and employ a combination of Maxwell and Hooke elements to describe their viscoelastic behaviour (Palmas and Rigato 1967, Kondo and Susset 1973, Regnier *et al* 1983). Despite their robustness and practicality, they do not capture the underlying physiological mechanisms of biomaterial behaviour and, therefore, have failed to integrate information about tissue composition and structure with their mechanical properties.

A knowledge of the mechanical properties of the tissue of the UB wall is crucial for the integration of motor functions into a biologically plausible biomechanical model. Most experiments on the bladder tissue under simple and complex loading protocols have been conducted on animals with only a few studies dedicated to the investigation of the human organ *per se* (Alexander 1971, 1976, Kondo *et al* 1972, Coolsaet *et al* 1975a, 1975b, 1976, van Mastright *et al* 1978, van Mastrigt and Nagtegaal 1981, Andersson *et al* 1989, Venegas *et al* 1991, Dahms *et al* 1998, Wagg and Fry 1999, Finkbeiner 1999, Sacks 2000, Gloeckner *et al* 2002, Gloeckner 2003,

doi:10.1088/978-0-7503-3567-6ch2

Korossis *et al* 2009, Nagatomi *et al* 2008, Parekh *et al* 2010, Wognum 2010, Martins *et al* 2011, Zanetti *et al* 2012, Chantereau *et al* 2014, Natali *et al* 2015, Borsdorf *et al* 2019). Linear strips for uniaxial stretching were usually excised from the organ. It was assumed that the muscle fibres were fully relaxed with the mechanical contribution attributed to mechanochemically inert components of smooth muscle cells alongside elastin and collagen fibres. *In vitro* quasi-static and dynamic tension tests were performed along two structurally defined orthogonal directions of anisotropy—the longitudinal (λ_l) and circumferential (λ_c). Their orientation coincided with the long and circumferential axes of the bladder, respectively. Assuming the homogeneity of the stress–strain fields and the incompressibility of the tissue, the passive force and stretch ratios ($T^{\mathrm{p}}_{c,\,l} - \lambda_{c,l}$) were calculated. The interpolation of data in the preferred axes of structural anisotropy yield

$$T^{\mathrm{p}}_{(c,\,l)} = c_1[\exp c_2(\lambda_{(c,l)} - 1) - 1], \quad \lambda_{(c,l)} > 1, \tag{2.1}$$

where c_1, c_2 are empirical mechanical constants.

Experimental results have demonstrated that the tissue has nonlinear, pseudoelastic properties and is similar to other biological materials. Analysis of the $T^{\mathrm{p}}_{c,\,l}(\lambda_{c,l})$ curves has shown a characteristic 'triphasic' response with a nonlinear transition between the low and high elastic states. Overall analysis across different species has revealed that the bladder wall has a considerable inherent inhomogeneity in its material properties and does not stretch equally in all directions being more compliant circumferentially than longitudinally. It is noteworthy that while insignificant differences between the loading and unloading curves are present due to 'biological hysteresis', the force-stretch ratio responses are independent of the stretching rate.

Histoarchitectural correlations with the dynamics of stress–strain development in the UB wall have revealed that the uncoiling of collagen fibres and small randomly oriented crack growth already begins at the early stages of bladder filling. These steadily increase in size as the bladder distends. There is a disrupture in the dense packaging of the fibrillary—collagen and elastin—matrix with the expansion and confluence of multiple small fractures. The distribution and orientation of elastin fibres in the bladder wall is both region and direction dependent. Most elastin is present in the ventral and lateral regions and appears to be oriented predominantly circumferentially. The detrusor muscle and collagen fibres are most compact within the lower body with the trigone regions being the least affected by distension.

Viscoelastic properties of the bladder wall tissue have been studied extensively on uniaxially loaded strips *in vitro*, and whole organ *in vivo*. The ramp and quasi-static loading protocols were employed in experimental settings. The quasi-linear viscoelastic model was used to describe the strain history dependence and hysteresis (Fung 1993). It assumed that the relaxation function $K(\lambda, t)$ is the product of the pseudoelastic response $T(\lambda)$ and a reduced relaxation function $G(t)$

$$K(\lambda, t) = T_0(\lambda) + \int_0^t T[\lambda(t - \tau)]\frac{\partial G(\tau)}{\partial \tau}\partial\tau \tag{2.2}$$

where

$$G(t) = \frac{1 + c_d[X(t/\tau_2) - X(t/\tau_1)]}{1 + c_d \ln(\tau_2/\tau_1)} \quad \text{for } \tau_1 \leqslant \tau \leqslant \tau_2, \tag{2.3}$$

and

$$X(t/\tau) = \int_0^\infty (e^{-t}/t)\mathrm{d}t \quad \text{where } (t/\tau) \leqslant \pi. \tag{2.4}$$

In the above, c_d is the decay parameter, and τ_1, τ_2 are the fast and slow time constants, respectively.

Results of stress relaxation studies have revealed indifference in biomaterial responses to quasi-static, ramp-and-hold and oscillatory modes of loading along the structural axes of anisotropy. There is a shift of the stiffness and damping curves towards the smaller frequencies of applied load and a decrease in the slope with higher stress levels indicating that larger stresses result in less relaxation with the damping more effective at smaller frequencies.

Biaxial tests to investigate, *in vitro*, the bladder wall tissue of different animals under quasi-static and dynamic loadings have been conducted on square-shaped specimens (Gloeckner *et al* 2002, Gloeckner 2003). These studies allow the deduction of full in-plane mechanical properties of the tissue. The edges of the specimens were aligned parallel and perpendicular to the orientation of the longitudinal and circular smooth muscle fibres. The experimental protocol to obtain force-stretch ratio curves $T^{\mathrm{p}}_{c,\,l}(\lambda_c, \lambda_l)$ used constant stretch ratios of $\lambda_l : \lambda_c$.

The in-plane passive $T^{\mathrm{p}}_{c,\,l}$ forces under biaxial loading are calculated as

$$T^{\mathrm{p}}_{c,\,l} = \frac{\partial \rho W}{\partial(\lambda_{c,l} - 1)}, \tag{2.5}$$

where the pseudo-strain energy density function W is chosen in the form

$$\rho W = \frac{1}{2}[c_3(\lambda_l - 1)^2 + 2c_4(\lambda_l - 1)(\lambda_c - 1) + c_4(\lambda_c - 1)^2 + \\ + c_6 \exp(c_7(\lambda_l - 1)^2 + c_8(\lambda_c - 1)^2 + 2c_9(\lambda_l - 1)(\lambda_c - 1))],$$

where ρ is the density of the undeformed tissue.

Bladders of pigs, rats and dogs under biaxial loading exhibit a complex response including nonlinear pseudoelasticity, transverse anisotropy and finite deformability with no dependence on the stretch rate. The curves, $T^{\mathrm{p}}_{c,\,l}(\lambda_c,\lambda_l)$, show that as the stretch ratio in one direction increases gradually, the extensibility along the other decreases. There is a concomitant increase in the stiffness of the biomaterial. The maximum force the tissue can bear during the biaxial tests depends on the ratio λ_l:λ_c. Experiments have shown that the shear force applied to the tissue is significantly less

$10^{-2}\,T^{\mathrm{p}}_{\mathrm{maxc},\,l}$ compared with the stretch force. Biaxial mechanical quasi-static loading combined with multiphoton microscopy imaging and immuno-histochemical staining of the bladder wall samples has demonstrated the sequential recruitment of collagen and elastin fibres to provide a correlation analysis between the morphological structure and mechanical characteristics of the tissue (Cheng *et al* 2018).

Investigations into the uniaxial and biaxial mechanical properties of actively contracting tissue remain a challenge. The main problem is to keep specimens physiologically viable and stable, i.e. for *in vitro* samples to reproduce myoelectrical patterns consistent with those observed *in vivo*. Seydewitz *et al* (2017) were the first to formulate a 3-dimensional continual mechano-electrochemical-coupled model of the contractile behaviour of the UB wall. Here the latter is treated as a multiphase biocomposite comprised of collagen, elastin fibres and ECM (passive part), and SMCs (contractile part). Following the conventional approach, the strain energy function is additively decomposed into

$$W = W_e(\mathbf{C}) + W_c(\mathbf{C}) + W_{\mathrm{ECM}}(\mathbf{C}) + W_{\mathrm{SM}}(\mathbf{V},\, \mathbf{C},\, [\mathrm{Ca}_i^{2+}]) \tag{2.6}$$

where the network of elastin fibres and the ground matrix satisfy the isotropic neo-Hookean material model, $W_{e(\mathrm{ECM})}(\mathbf{C}) = \frac{\mu_{e(\mathrm{ECM})}}{2}(\mathrm{tr}\mathbf{C} - 3)$; the anisotropic nonlinear behaviour of collagen fibres is described by $W_c(\mathbf{C}) = \sum_i^{n_c} f_{c,i}\, T_i^{\mathrm{p}}$; and for SMCs: $W_{\mathrm{SM}}(\mathbf{V},\, \mathbf{C},\, [\mathrm{Ca}_i^{2+}]) = \sum_j^{n_{SM}} f_{\mathrm{SM},j}(T^{\mathrm{a}}_{\mathrm{SM},\,j} + T^{\mathrm{p}}_{\mathrm{SM},\,j})$. In the above, $\mu_{e(\mathrm{ECM})}$ are the shear moduli for elastin (subscripts e) and ECM, $\mathbf{C}$ is the right strain Cauchy–Green tensor, $f_{c,i}$, $f_{\mathrm{SM},j}$ are the fractions of collagen and smooth muscle fibres aligned in the directions i and j, respectively, $n_{c(SM)}$ is the total number of incorporated directions, $\mathbf{V}$ is the electrical potential, $[\mathrm{Ca}_i^{2+}]$ is the intracellular calcium concentration, and $T^{\mathrm{a}}_{\mathrm{SM},\,j}$, $T^{\mathrm{p}}_{c,\,\mathrm{SM},\,j(i)}$ are the experimentally obtained and approximated functions in the form (Holzapfel 2001)

$$\begin{gathered} T^{\mathrm{p}}_{c,\mathrm{SM},(i,j)} = \begin{cases} \tilde{c}_1\left[\exp \tilde{c}_2\left(\lambda^2_{c,\mathrm{SM},(i,j)} - 1\right)^2 - 1\right], & \lambda_{c,\mathrm{SM},(i,j)} \\ 0, & \text{else} \end{cases} \\ T^{\mathrm{a}}_{\mathrm{SM},j} = \psi P_{\mathrm{opt}} \int \exp\left[\frac{-(\lambda_{\mathrm{SM},j} - \lambda_{\mathrm{opt}})^2}{2\xi^2_{\mathrm{SM}}}\right] d\lambda_{\mathrm{SM},j}. \end{gathered} \tag{2.7}$$

Here, the contraction processes satisfy the four-state, cross-bridge model of Hai and Murphy (1988) with ψ referring to the chemical degree of activation, λ_{opt}, P_{opt}- the optimum stretch and the corresponding generated stress, respectively, and ξ_{SM} being the empirical parameter. The dynamics of the electrical potential is described by the two-variable FitzHugh–Nagumo-type model (FitzHugh 1961, Nagumo *et al* 1962). In this model, parameters are determined experimentally during uniaxial loadings of strips excised from porcine bladders in the orientation directions of SM fibres.

At present it is practically impossible to evaluate biaxial active stress–strain states of UB wall tissue owing to the inability to sustain and control simultaneously the spiking and contractile activity of SM syncytia.

2.2 Multiphase models

The first detailed research to develop a multiphase structural constitutive model for soft tissue with applications to the mechanics of the UB has been undertaken by Wognum (2010). Assuming that: (i) the soft tissue is an idealised network of muscle and undulated (in an undeformed state) collagen fibres embedded into a compliant ground matrix, (ii) the mechanical net response is the sum of responses of individual fibres, (iii) the tissue is incompressible, (iv) the strain energy density function W is the composition

$$W = \phi_{\mathrm{ECM}} W_{\mathrm{ECM}}(\mathbf{E}) + \phi_{\mathrm{SM}} W_{\mathrm{SM}}(\mathbf{E}), \tag{2.8}$$

where $\phi_{\mathrm{ECM}}, \phi_{\mathrm{SM}}$ are the ECM and SM volume fractions, respectively, and E is the Green-Lagrange strain tensor, the expression for the Piola–Kirchhoff stress S is obtained as

$$\mathbf{S}(\mathbf{E}) = \frac{\partial W}{\partial \mathbf{E}} - l_{\mathrm{m}}\mathbf{C}^{-1} = \phi_{\mathrm{ECM}}\mathbf{S}_{\mathrm{ECM}}(\mathbf{E}_{\mathrm{ECM}}) + \phi_{\mathrm{SM}}\mathbf{S}_{\mathrm{SM}}(\mathbf{E}_{\mathrm{SM}}) - l_{\mathrm{m}}\mathbf{C}^{-1}. \tag{2.9}$$

Here, l_{m} is the Lagrange multiplier and $\mathbf{C}$ is the right Cauchy–Green strain tensor. The subsequent recruitment of fibres during loading in an ensemble of weight bearing elements suggests that the complete stress in the ECM and SMs is

$$\mathbf{S}_{\mathrm{i}}(\mathbf{E}_{\mathrm{i}}) = \eta\phi_{\mathrm{f}} \int_{-\pi/2}^{\pi/2} R(\hat{\theta}) \left\{ \int_{0}^{E_{ens}} D(x) E_{\mathrm{ens}}(\hat{\theta}) dx \right\} \bar{\mathbf{r}} \otimes \bar{\mathbf{r}} d\hat{\theta} \quad i = \mathrm{ECM, SM}. \tag{2.10}$$

Here, the parameters and functions are referred to the collagen/smooth muscle fibre: η is the modulus, ϕ_{f} is the volumetric fraction, $R(\hat{\theta})$ is the distribution function, $\hat{\theta}$ is the orientation angle in the undeformed configuration, Eens is the fibre ensemble slack strain, $D(x)$ is the recruitment function, $\bar{\mathbf{r}}$ is the orientation vector. The exact forms of $R(\hat{\theta})$ and $D(x)$ are assumed and given *a priori*.

The formulation of a fibre-reinforced viscohyperelastic constitutive model of the bladder wall has been put forward by Natali *et al* (2015). Operating within the general framework of soft tissue mechanics, the specific hyperelastic potential is decoupled as

$$W^0(\mathbf{C}) = W_m^0(\mathbf{C}) + W_f^0(\mathbf{C}, \mathbf{a}_0, \mathbf{b}_0) \tag{2.11}$$

where $\mathbf{C}$ is the right Cauchy–Green strain tensor, $\mathbf{a}_0$, $\mathbf{b}_0$ are the longitudinal and transverse directional vectors, and $W_m^0(\mathbf{C})$, $W_f^0(\mathbf{C}, \mathbf{a}_0, \mathbf{b}_0)$ are the ground matrix (subscript m) and connective tissue fibre (subscript f) potentials, respectively, given by

$$\begin{aligned} W_m^0(\mathbf{C}) &= -p(I_3 - 1) + \frac{G_1}{\alpha_1}\exp(\alpha_1(I_1 - 3) - 1) \\ W_f^0(\mathbf{C}, \mathbf{a}_0, \mathbf{b}_0) &= W_{fAB}^0(I_4) + W_{fT}^0(I_6). \end{aligned} \tag{2.12}$$

In the above, $I_1 = \text{tr}(\det(\mathbf{C})^{-1/3}\mathbf{C})$, I_4, I_6 are structural invariants that reflect tissue stretching along $\mathbf{a}_0$ and $\mathbf{b}_0$, respectively, p is the Lagrange multiplier, G_1 is the tissue shear stiffness, α_1 is the empirical parameter, and $W^0_{fAB}(I_4)$, $W^0_{fT}(I_6)$ are given by (Natali *et al* 2012)

$$W^0_{fAB(T)}(I_{4(6)}) = \frac{C_{4(6)}}{\alpha^2_{4(6)}}[\exp(\alpha_{4(6)}(I_{4(6)} - 1) - \alpha_{4(6)}(I_{4(6)} - 1) - 1]. \tag{2.13}$$

Here, $C_{4(6)}$ are the fibres initial stiffness, and $\alpha_{4(6)}$ describe the stiffening of fibres with a stretch.

The first Piola–Kirchhoff nominal hyperelastic stress tensor contributions are calculated as (Natali *et al* 2010)

$$\mathbf{P}^0_m = 2\mathbf{F}\frac{\partial W^0_m}{\partial \mathbf{C}} = -p\mathbf{F}^{-1} + G_1 \exp(\alpha_1(\tilde{I}_1 - 3))(2J^{-2/3}\mathbf{F} - 2/3\tilde{I}_1\mathbf{F}^{-T})$$

$$\begin{aligned} \mathbf{P}^0_{fAB}(I_4) &= 2\mathbf{F}\frac{\partial W^0_{fAB}}{\partial \mathbf{C}} = \frac{2C_4}{\alpha^2_4}[\exp(\alpha_4(I_4 - 1) - 1]\mathbf{F}(\mathbf{a}_0 \otimes \mathbf{a}_0) \\ \mathbf{P}^0_{fT}(I_6) &= 2\mathbf{F}\frac{\partial W^0_{fT}}{\partial \mathbf{C}} = \frac{2C_6}{\alpha^2_6}[\exp(\alpha_6(I_6 - 1) - 1]\mathbf{F}(\mathbf{b}_0 \otimes \mathbf{b}_0) \end{aligned} \tag{2.14}$$

and the evolution of viscous variables, $\mathbf{q}_{(m,fAB,fT)}$, to quantify the relaxation of stress during loading are obtained as a solution to

$$\frac{d}{dt}\mathbf{q}_i + \frac{1}{\tau_i}\mathbf{q}_i = \frac{\gamma_i}{\tau_i}\mathbf{P}^0_i, \qquad i = m, fAB, fT \tag{2.15}$$

where τ_i are relaxation times, γ_i are the relative tissue stiffness, and $\mathbf{F}$ is the deformation gradient.

The constitutive relationship for the UB wall tissue in its final form is (Natalli *et al* 2015)

$$\mathbf{P}(\mathbf{C}, \mathbf{q}_i) = \sum_{i=m,fAB,fT} (\mathbf{P}^0_i(\mathbf{C}) - \mathbf{q}_i). \tag{2.16}$$

The parameters and constants in the model have been evaluated from *in vitro* studies conducted on samples excised from a pig UB. The results of theoretical predictions show excellent agreement with the curves obtained experimentally.

The general nature of this approach, the decoupling of the hyperelastic potential, has the potential to explain the underlying remodelling mechanisms of individual constituents, i.e. the process of uncoiling, straightening and reorientation of fibres along the direction of the applied force under normal physiological conditions, and to estimate their role in various pathologies.

Recent technological advancements in bioengineering have brought new opportunities to bladder wall tissue modelling and reconstruction. Approaches are based on the use of porous polymer, e.g. polyurethane-poly-lactic-co-glycolic acid, ε-caprolactone, polyvinyl alcohol, nanofibrous poly-L-lactide, and/or cell-seeded matrices, e.g. porcine extracellular, type I collagen, porcine bladder and small intestine submucosa, autologous vesical equivalent, as scaffolds in which tissue regeneration is promoted (Shakhssalim *et al* 2017, Tiemessen *et al* 2017, Yao *et al* 2013, Chen *et al* 2012, Bouhout *et al* 2011, Davis *et al* 2011, Tian *et al* 2010, Shaikh *et al* 2010, Serrano-Aroca *et al* 2018, Liu *et al* 2009). Biological scaffolds used have a honeycomb structure with a hexagonal unit cell (core) as a periodic microstructure. It replicates the real architecture of the supporting connective stroma of the *tunica muscularis* of the UB (see chapter 1). The inner, *tunica mucosa*, and *submucosa*, and outer, *tunica serosa*, layers serve as face sheets for the core. The variable thickness of collagen and elastin fibres, their complex intertwining, the nodal angulation, the pattern and size of fenestrae, the presence of the ground matrix, smooth muscle and other cells guarantee high compliance, deformability and low bending rigidity of the tissue. However, at the time of writing no mathematical models of collagen and elastin based scaffolds have been proposed.

References

Alexander R S 1971 Mechanical properties of the urinary bladder *Am. J. Physiol.* **220** 1413–21

Alexander R S 1976 Series elasticity of urinary bladder smooth muscle *Am. J. Physiol.* **231** 1337–42

Andersson K E, Kronström A and Bjerle P 1989 Viscoelastic properties of the normal human bladder *Scand. J. Urol. Nephrol.* **23** 115–20

Borsdorf M, Tomalka A, Stutzig N, Morales-Orcajo E, Böl M and Siebert T 2019 Locational and directional dependencies of smooth muscle properties in pig urinary bladder *Front. Physiol.* **10** 63

Bouhout S, Gauvin R, Gibot L, Aubé D and Bolduc S 2011 Bladder substitute reconstructed in a physiological pressure environment *J. Pediatr. Urol.* **7** 276–82

Chantereau P, Brieu M, Kammal M, Farthmann J, Gabriel B and Cosson M 2014 Mechanical properties of pelvic soft tissue of young women and impact of aging *Int. Urogynecol. J.* **25** 1547–53

Chen L, Wei T, Wang Y, Zhang J, Li H and Wang K 2012 Simulated bladder pressure stimulates human bladder smooth muscle cell proliferation via the PI3K/SGK1 signaling pathway *J. Urol.* **188** 661–7

Cheng F, Birder L, Kullmann F A, Hornsby J, Watton P, Watkins S C, Thompson M and Robertson A M 2018 Layer dependent role of collagen recruitment during loading of the rat bladder wall *Biomech. Model. Mechanobiol.* **17** 403–17

Coolsaet B L R A, van Duyl W A, van Mastrigt R and Schouten J W 1975a Visco-elastic properties of bladder wall strips *Invest. Urol.* **12** 351–5

Coolsaet B L R A, van Duyl W A, van Mastrigt R and van der Zwart A 1975b Visco-elastic properties of the bladder wall *Urol. Int.* **30** 16–26

Coolsaet B L R A, van Mastrigt R, van Duyl W A and Huygen R E 1976 Viscoelastic properties of bladder wall strips at constant elongation *Invest. Urol.* **13** 435–40

Dahms S E, Piechota H J, Dahiya R, Lue T F and Tanagho E A 1998 Composition and biomechanical properties of the bladder acellular matrix graft: comparative analysis in rat, pig and human *Br. J. Urol.* **82** 411–9

Davis N F, mooney R, Piterina A V, Callanan A, McGuire B B, Flood H D and McGloughlin T M 2011 Construction and evaluation of urinary bladder bioreactor for urologic tissue-engineering purposes *Urology* **78** 954–60

Finkbeiner A E 1999 *In vitro* responses of detrusor smooth muscle to stretch and relaxation *Scand. J. Urol. Nephrol. Suppl.* **201** 5–11

Fitzhugh R 1961 Impulses and physiological states in theoretical models of nerve membrane *Biophysics* **J1** 445–66

Fung Y C 1993 *Biomechanics: Mechanical Properties of Living Tissues* 2nd edn (New York: Springer)

Gloeckner D C 2003 Tissue biomechanics of the urinary bladder wall *Dr Phil Thesis* University Pittsburgh

Gloeckner D C, Sacks M S, Fraser M O, Somogyi G T, de Groat W C and Chancellor M B 2002 Passive biaxial mechanical properties of the rat bladder wall after spinal cord injury *J. Urol.* **167** 247–52

Hai C M and Murphy R A 1988 Cross-bridge phosphorylation and regulation of latch state in smooth muscle *Am. J. Physiol. Cell. Physiol.* **254** C99–106

Holzapfel G A 2001 Biomechanics of soft tissue *Handbook of Mater. Beh. Models* **3** 1049–63

Kondo A and Susset J G 1973 Physical properties of the urinary bladder detrusor muscle. A Mechanical model based upon the analysis of stress relaxation curve *J. Biomech.* **6** 141–51

Kondo A, Susset J G and Lefaivre J 1972 Viscoelastic properties of bladder I. Mechanical model and its mathematical analysis *Invest. Urol.* **10** 154–63

Korossis S, Bolland F, Southgate J, Ingham E and Fisher J 2009 Regional biomechanical and histological characterization of the passive porcine urinary bladder: implications for augmentation and tissue engineering strategies *Biomaterials* **30** 266–75

Liu Y, Bharadwaj S, Lee S J, Atala A and Zhang Y 2009 Optimization of a natural collagen scaffold to aid cell-matrix penetration for urologic tissue engineering *Biomaterials* **30** 3865–73

Martins P A L S, Filho A L S, Fonseca A M R M, Santos A A, Santos L, Mascarenhas T, Jorge R M N and Ferreiraet A J M 2011 Uniaxial mechanical behavior of the human female bladder *Int. Urogynecol. J.* **22** 991–5

Nagatomi J, Toosi K K, Chancellor M B and Sacks M S 2008 Contribution of the extracellular matrix to the viscoelastic behaviour of the urinary bladder wall *Biomech. Model. Mechanobiol.* **7** 395–404

Nagumo J, Arimoto S and Yoshizawa S 1962 An active pulse transmission line simulating nerve axon *Proc. IRE* **50** 2061–70

Natali A N, Audenino A L, Artibani W, Fontanella C G, Carniel E L and Zanetti E M 2015 Bladder tissue biomechanical behavior: Experimental tests and constitutive formulation *J. Biomech.* **48** 3088–96

Natali A N, Fontanella C G and Carniel E L 2010 Constitutive formulation and analysis of heel tissue pad mechanics *Med. Eng. Phys.* **32** 516–22

Natali A N, Fontanella C G and Carniel E L 2012 A numerical model for investigating the mechanics of calcaneal fat pad region *J. Mech. Behav. Biomed. Mater.* **5** 216–23

Palmas G and Rigato M 1967 Confronto fra il compotamento viscoelastico della vescica urinaria e quello di un modello meccanico ideale *Atti. Acad. Fisiocrit. Siena. Med. Fis.* **16** 1731–63

Parekh A, Cigan A D, Wognum S, Heise R L and Chancellor M B 2010 *Ex vivo* deformations of the urinary bladder during whole bladder filling: contributions of extracellular matrix and smooth muscle *J. Biomech.* **43** 1708–16

Regnier C, Kolsky H, Richardson P D, Ghoinem G M and Susset J G 1983 The elastic behavior of the urinary bladder for large deformations *J. Biomech.* **16** 915–22

Sacks M S 2000 Biaxial mechanical evaluation of planar biological materials *J. Elastic.* **61** 199–246

Serrano-Aroca A, Vera-Donoso C and Moreno-Manzano V 2018 Bioengineering approaches for bladder regeneration *Int. J. Mol. Sci.* **19** 1796

Seydewitz R, Menzel R, Siebert T and Böl M 2017 Three-dimensional mechano-electrochemical model for smooth muscle contraction of the urinary bladder *J. Mech. Behav. Biomed. Mater.* **75** 128–46

Shaikh F M, O'Brien T P, Callanan A, Kavanagh E G, Burke P E, Grace P A and McGloughlin T M 2010 New pulsatile hydrostatic pressure bioreactor for vascular tissue-engineered constructs *Artif. Organs* **34** 153–8

Shakhssalim N, Soleimani M, Dehghan M M, Rasouli J, Taghizadeh-Jahed M, Torbati P M and Naji M 2017 Bladder smooth muscle cells on electrospun poly(ε -caprolactone)/poly(l-lactic acid) scaffold promote bladder regeneration in a canine model *Mater. Sci. Eng.* **75** 877–84

Tian H, Bharadwaj S, Liu Y, Ma H, Ma P X, Atala A and Zhang Y 2010 Myogenic differentiation of human bone marrow mesenchymal stem cells on a 3D nano fibrous scaffold for bladder tissue engineering *Biomaterials* **31** 870–7

Tiemessen D, de Jonge P, Daamen W, Feitz W, Geutjes P and Oosterwijk E 2017 The effect of a cyclic uniaxial strain on urinary bladder cells *World J. Urol.* **35** 1531–9

van Mastright, Coolsaet B L R A and van Duyl W A 1978 Passive properties of the urinary bladder in the collection phase *Med. Biol. Eng. Comput.* **16** 471–82

van Mastrigt R and Nagtegaal J C 1981 Dependence of the viscoelastic response of the urinary bladder wall on strain rate *Med. Biol. Eng. Comput.* **19** 291–6

Venegas J G, Woll J P, Woolfson S B, Cravalho E G, Resnick N and Yalla S V 1991 Viscoelastic properties of the contracting detrusor II: Experimental approach *Am. J. Physiol.* **261** 364–75

Wagg A and Fry C H 1999 Viscoelastic properties of isolated detrusor smooth muscle *Scand. J. Urol. Nephrol. Suppl.* **201** 12–8

Wognum S 2010 A multi-phase structural constitutive model for insights into soft tissue remodelling mechanisms *Dr Thesis* University Pittsburgh

Yao C, Hedrick M, Pareek G, Renzulli J, Haleblian G and Webster T J 2013 Nanostructured polyurethane-poly-lactic-co-glycolic acid scaffolds increase bladder tissue regeneration: An *in vivo* study *Int. J. Nanomed.* **8** 3285–96

Zanetti E M, Perrini M, Bignardi C and Audenino A L 2012 Bladder tissue passive response to monotonic and cyclic loading *Biorheology* **49** 49–63

IOP Publishing

The Brain–Bladder Axis in Tissue Growth and Remodelling

Roustem N Miftahof and Christian J Cyron

Chapter 3

Models of the urinary bladder

Strength lies in differences, not in similarities.

— Stephen R Covey

The systems biology approach simulating the urinary bladder as a multihierarchical biologically plausible entity is analysed. Currently existing models integrating biomechanical, urodynamic and regulatory components are described. A model of the urinary bladder as a soft biologically active shell is formulated.

3.1 The bladder as shell structure

Publications on modelling the biomechanics of the urinary bladder (UB) as an organ are scarce. The primary focus has been on modelling the UB as a reservoir (Regnier *et al* 1983, van Beek 1997, Vlastelica *et al* 2007, Spirka and Damaser 2007, Korkmaz and Rogg 2007, Fry *et al* 2011), its neuronal control mechanisms (Hosein and Griffiths 1990, Kinder *et al* 1999, Bastianssen *et al* 1996a, 1996b, van Duin *et al* 2000, Fernández *et al* 2004, Pérez *et al* 2008), and simulating urine–bladder–urethra interactions during filling and micturition (Damaser and Lehman 1993, Fletcher *et al* 1997, Damaser 1999, Křen *et al* 2001, Celik *et al* 2007, Zang *et al* 2010, Kim *et al* 2011).

The urinary bladder has been typically treated as a thin elastic shell subjected to external and internal quasi-static loads. Initially, it was suggested that the organ could be approximated by a simple shape—a sphere. Assuming incompressibility, homogeneity, mechanical isotropy and physical linearity of the tissue, Regnier *et al* (1983) established the dependence of inflation pressure on extension ratios. Damaser and Lehman (1993) investigated the role of more complex bladder shapes—the prolate and oblate spheroids—on the intravesicular pressure–volume $p(V)$ response during filling. Results showed that the $p(V)$ curves obtained for prolate and oblate spheroids for a wide range of eccentricity, $0.1 < \varepsilon < 0.9$, did not deviate significantly from the sphere. Later, Damaser (1999) incorporated low muscle tone in the model and reconfirmed his earlier findings. An inverse problem to determine the

doi:10.1088/978-0-7503-3567-6ch3

stress–strain relationships of the UB wall was considered by Korkmaz and Rogg (2007). Based on the assumption that the organ is a thin spherical shell subjected to complex loading, the wall tension was calculated as a function of the intra-abdominal pressure.

In van Beek's model (1997) of the UB as a thick-walled sphere, special emphasis was given to the evaluation of the effect of muscle fibre orientation and physical nonlinearity of the tissue on the dynamics of organ deformation. To simulate active stress, the classical Hill model composed of a series of passive elastic and contractile elements was used. The artificial activation, deactivation and (the) length dependence functions were introduced to mimic the effect of regulatory mechanisms. Three different geometries of fibre winding were studied: the circumferential, longitudinal and oblique. Results of numerical simulations in the case of circular winding of muscle fibres demonstrated a gradual change in the spherical configuration of the shell (bladder) to a prolate spheroid during the collection phase, with a reversal to its initial state (at the end of void) when voiding is complete. Analysis of stress distribution showed the highest stress values at the inner wall in the equator region and the lowest at the poles. A bladder with longitudinal fibre geometry attained the shape of an oblate spheroid at the end of filling, whilst the pattern of stress–strain distribution in the case of circumferential geometry was inverse.

A conceptually analogous approach was employed by Vlastelica *et al* (2007) in their numerical modelling of the UB as an axisymmetric thin biologically active shell. The wall tissue was assumed to be homogenous and to possess nonlinear elastic properties. However, the supposition is valid for 'passive' biomaterials only and becomes invalid when any muscle activity is present or suspected. Simulation results of the passive relaxation and detrusor activation phases show maximum circumferential stress occurring at the lower part of the bladder and at the fundus. The body of the organ experiences a uniform axial and circumferential stress distribution throughout.

The first plausible system biology approach to simulate the UB as a multi-hierarchical system integrating the neural network with biomechanical and urodynamic components in one biologically plausible quantitative myocybernetic model was undertaken by Bastianssen *et al* (1996a, 1996b). The organ was considered as a thin spherical isotropic shell and the sphincter, a cylinder. The dynamics of neural excitatory and inhibitory stimuli that trigger detrusor and urethral sphincter contraction and relaxation, respectively, are described by an *a priori* definition of normalised activity functions. Based on the assumption that the detrusor undergoes quasi-static isometric contractions, it was therefore treated as a homogenous incompressible viscoelastic continuum. The total actual tensile stresses in the muscle (D) and sphincter (S) were decomposed into a sum of (the) active and passive stresses

$$\begin{aligned} \sigma_{\mathrm{D}} &= k_{\mathrm{area}}\big(f_{\mathrm{aD}}\sigma_{\max}\sigma_{\mathrm{v}}(v_{\mathrm{D}})\sigma_{\mathrm{l}}(l_{\mathrm{D}}) + \sigma_{\mathrm{e}}(l_{\mathrm{D}}) + \sigma_{\mathrm{ve}}(v_{\mathrm{D}})\big) \\ \sigma_{\mathrm{S}} &= k_{\mathrm{thick}}\big(f_{\mathrm{aS}}\sigma_{\max}\sigma_{\mathrm{v}}(v_{\mathrm{S}})\sigma_{\mathrm{l}}(l_{\mathrm{S}}) + \sigma_{\mathrm{e}}(l_{\mathrm{S}})\big) \end{aligned} . \tag{3.1}$$

Here, $\sigma_{\max}$, $\sigma_{\mathrm{l}}(l_{\mathrm{D}})$, $\sigma_{\mathrm{l}}(l_{\mathrm{S}})$ are the maximal and actual isometric stresses of the detrusor and urethral sphincter, respectively, $\sigma_{\mathrm{v}}(v_{\mathrm{D}})$, $\sigma_{\mathrm{v}}(v_{\mathrm{S}})$ are the velocity-dependent stresses obtained at an optimal length of muscle fibres that satisfy the Hill equation, $\sigma_{\mathrm{e}}(l_{\mathrm{D}})$, $\sigma_{\mathrm{e}}(l_{\mathrm{S}})$, $\sigma_{\mathrm{ve}}(v_{\mathrm{D}})$ are the elastic and viscoelastic stresses, k_{area}, k_{thick} are the ratio of the cross-sectional area of the detrusor taken at optimal muscle fibre length to the actual area and thickness of the sphincter, respectively.

In the model, most parameters have both physiological and physical meanings. Therefore, it was possible for the authors to define a range for variation and measure their actual values. Despite its mathematical and biological *naiveté*, the model results resemble the traces obtained from clinical urodynamic studies. Several other models of the lower urinary tract have been proposed which are, to a certain extent, variations of the Bastiansen *et al* model.

Argungu *et al* (2018) studied the UB during filling and attempted to analyse simultaneously the mechanosensory response in the afferent nerve endings. The organ has been treated as an isotropic spherical shell subjected to the uniformly distributed internal, $p_i(t)$, and external, $p_e(p_e = \text{const})$, radial pressure. Deformations and displacements of the shell are finite and symmetric. The biomaterial is incompressible and possesses quasi-linear viscoelastic characteristics with the time-dependent stress Cauchy tensor, $\mathbf{T}(t)$, given by (De Pascalis *et al* 2014)

$$\mathbf{T}(t) = \mathbf{F}(t)(\mathbf{\Pi}^e_D(t) + \int_0^t D'(t-s)\mathbf{\Pi}^e_D(s)ds)\mathbf{F}^{\mathrm{T}}(t) - p(R,t)\mathbf{I}, \tag{3.2}$$

where $\mathbf{I}$ is the identity tensor, $D(t)$ is a relaxation function where the prime sign denotes the derivative with respect to s, $p(R,t)$ is a Lagrange multiplier associated with the incompressibility constraint, R is the radial coordinate of the shell, and $\mathbf{\Pi}^e_D(t)$ is the deviator of the elastic Piola–Kirchhoff stress tensor defined by

$$\mathbf{\Pi}^e_D(t) = \mu\left(\mathbf{I} - \frac{I_1\mathbf{C}^{-1}}{3}\right). \tag{3.3}$$

Here

$$I_1 = \frac{R^4}{(R^3+\alpha(t))^{4/3}} + 2\frac{(R^3+\alpha(t))^{2/3}}{R^2},$$

and $\mathbf{C}$ is the right Cauchy–Green stress tensor, $\alpha(t)$ is the deformation parameter, μ is the material constant. Explicit expression for the pressure difference across the wall of the bladder has been obtained

$$\begin{aligned} p_i(t) - p_e = \int_A^B \Bigg(& \rho\frac{R^2}{R^3+\alpha(t)}\frac{\partial^2 U}{\partial t^2} + 2\mu\Bigg(\frac{(R^3+\alpha(t))^{1/3}}{R^2} - \frac{R^2}{(R^3+\alpha(t))^{7/3}}\Bigg) + \\ & + \frac{(R^3+\alpha(t))^{1/3}}{R^2}\int_0^t D'(t-s)\left(1 - \frac{I_1}{3}\frac{R^4}{(R^3+\alpha(s))^{4/3}}\right)ds - \\ & - \frac{R^6}{(R^3+\alpha(t))^{7/3}}\int_0^t D'(t-s)\left(1 - \frac{I_1}{3}\frac{(R^3+\alpha(s))^{4/3}}{R^4}\right)ds\Bigg)dR \end{aligned} \tag{3.4}$$

where the displacement $U(t) = r(t) - R = (R^3 + \alpha(t))^{1/3} - R$. For a particular choice of the relaxation function, i.e., the first term of the Prony series (De Pascalis *et al* 2014),

$$D(t) = \frac{\mu_\infty}{\mu} + \left(1 - \frac{\mu_\infty}{\mu}\right)\exp(-t/\tau),$$

the deformation parameter

$$\alpha(t) = 182\tanh(0.03(t - 320)),$$

and other constants (in the above, μ_∞ is the long-time infinitesimal shear modulus and τ is the relaxation time), the numerical results of intraluminal pressure and UB volume changes are obtained.

The modified Fitzhugh–Nagumo model has been employed to simulate the mechanosensory part of the model and the development of the action potential in the afferent nerve fibre (McKean 1970)

$$\begin{aligned} \frac{dv}{dt} &= -v + H(v - a) - y + \underline{g_M(\lambda(t))(v - v_M)} \\ \frac{dy}{dt} &= \varepsilon(v - d - cy). \end{aligned} \tag{3.5}$$

Here v, y are state variables, a, d, c, and ε are constants, and the underlined term describes the stretch activated current. A specific membrane conductance, $g_M(\lambda(t))$, satisfies the condition

$$g_M(\lambda(t)) = \begin{cases} \dfrac{g_m}{1 + k\exp(\lambda_c - \lambda)} & \text{if } t < t_{\text{empty}} \\ 0 & \text{if } t > t_{\text{empty}}, \end{cases} \tag{3.6}$$

where g_m is the maximal conductance, λ_c, λ are the maximum and current stretch of the shell, and k is the empirical variable. Although λ is supposed to be calculated from the mechanical part of the model of the bladder, authors of this experiment/proposition did not provide any link between the two.

3.2 The bladder as soft bioshell

3.2.1 General postulates

From the perspective of continuum mechanics, the assessment of experimental data on the UB as an anatomical entity and its tissue structure (constructive material) shows that:

(i) the wall of the bladder is formed by connective (collagen and elastin) and SM muscle fibres embedded in the ground substance matrix; the fibres are aligned along the axial and circumferential directions (and sustain) sustaining their orientation throughout loading/unloading of the UB

(ii) $\max(h/R) \leqslant 1/20$, where h is the thickness of the bladder wall ranging within 3.3 ± 1.1 (mm), R is the characteristic radius of the curvature varying between 8 to 10 (cm)
(iii) $h/L \sim O$, here L is the height that measures 10 cm in a fully distended bladder
(iv) the detrusor muscle possesses zero bending rigidity and does not withstand compression forces
(v) the actual configuration is dependent entirely on internal/ external loads distributed over the unit surface area, and
(vi) the stress–strain state is fully described by in-plane membrane forces per unit length.

From the above, it becomes clear that the UB as a mechanical system fulfills all the hypotheses and requirements of the theory of thin soft shells. Therefore, the general concepts and mathematical formulations of the theory can be rightfully used to model the organ. The fundamental difference between soft shells used in industrial engineering and that under current consideration (here) is that the latter generates active forces of contraction-relaxation as a result of intrinsic biological reactions. To emphasise the significance of this difference, the UB is referred to in the future as a thin soft biological shell or a bioshell for short (Miftahof and Akhmadeev 2019).

3.2.2 Governing equations in general curvilinear coordinates

Let $\Delta s^{(*)}$ and $\Delta m^{(*)}$ be the surface area and mass of a differential element of the soft bioshell in undeformed and deformed (superscript $*$) configurations. The position of a point $M^* \in S^*$ at any moment of time t is given by vector $\bar{r}(\alpha^1, \alpha^2, t)$. Densities, $\rho^{(*)}$, of the material in undeformed and deformed states are defined by

$$\rho^{(*)} = \lim_{\Delta s \to 0} \frac{\Delta m^{(*)}}{\Delta s^{(*)}} = \frac{dm^{(*)}}{ds^{(*)}}, \tag{3.7}$$

where

$$ds^{(*)} = \sqrt{a^{(*)}}\, d\alpha^1 d\alpha^2.$$

Applying the law of conservation of the mass to equation (3.2), we obtain

$$dm = dm^{(*)} = \rho^{(*)} ds^{(*)} = \rho^{(*)} \sqrt{a^{(*)}}\, d\alpha^1 d\alpha^2 = \rho \sqrt{a^{(*)}}\, d\alpha^1 d\alpha^2,$$

from where it follows: $\rho^{(*)} = \rho\sqrt{a/a^{(*)}}$.

Let $\bar{P}_s(\alpha_1, \alpha_2, t)$ be the resultant of the external, $\bar{P}_{(+)}(\alpha_1, \alpha_2, t)$, and internal, $\bar{P}_{(-)}(\alpha_1, \alpha_2, t)$, forces distributed over the outer and inner surfaces of the bioshell

$$\bar{P}_s(\alpha_1, \alpha_2, t) = \bar{P}_{(+)}(\alpha_1, \alpha_2, t) + \bar{P}_{(-)}(\alpha_1, \alpha_2, t).$$

The density of the resultant force per unit area of a deformed element $\bar{P}_s$ is defined by

$$\bar{p}(\alpha_1, \alpha_2, t) = \lim_{\Delta s^* \to 0} \frac{\bar{P}_s}{\Delta s^{(*)}}. \tag{3.8}$$

Similarly, we introduce the density of the mass force $\bar{F}(\alpha^1, \alpha^2, t)$ by

$$\bar{f}(\alpha^1, \alpha^2, t) = \lim_{\Delta m \to 0} \frac{\bar{F}}{\Delta m} = \frac{d\bar{F}}{dm} = \frac{1}{\rho}\frac{d\bar{F}}{ds}. \tag{3.9}$$

The resultant stress vectors $\bar{R}_i$ acting upon the differential element are

$$\begin{aligned}
&\bar{R}_1 = -(T^{11}\bar{e}_1 + T^{12}\bar{e}_2)\sqrt{a_{22}^*}\,d\alpha^2, \quad \bar{R}_2 = -(T^{21}\bar{e}_1 + T^{22}\bar{e}_2)\sqrt{a_{11}^*}\,d\alpha^1\\
&-\left(\bar{R}_1 + \frac{\partial \bar{R}_1}{\partial \alpha^1}d\alpha^1\right) = -(T^{11}\bar{e}_1 + T^{12}\bar{e}_2)\sqrt{a_{22}^*}\,d\alpha^2\\
&\qquad - \frac{\partial}{\partial \alpha^1}(T^{11}\bar{e}_1 + T^{12}\bar{e}_2)\sqrt{a_{22}^*}\,d\alpha^1 d\alpha^2\\
&-\left(\bar{R}_2 + \frac{\partial \bar{R}_2}{\partial \alpha^2}d\alpha^2\right) = -(T^{21}\bar{e}_1 + T^{22}\bar{e}_2)\sqrt{a_{11}^*}\,d\alpha^1\\
&\qquad - \frac{\partial}{\partial \alpha^2}(T^{21}\bar{e}_1 + T^{22}\bar{e}_2)\sqrt{a_{11}^*}\,d\alpha^1 d\alpha^2
\end{aligned} \tag{3.10}$$

where T^{ij} are contravariant components of the in-plane forces per unit length, $\bar{e}_i$ $(i = 1, 2)$ are the unit vectors $(\bar{e}_i = \bar{r}_i/|\bar{r}_i|)$ along the coordinate lines $(\alpha_1, \alpha_2) \in S$. Applying the law of conservation of momentum to equations (3.3)–(3.5), for the equation of motion of the soft shell in the vector form we obtain

$$\rho_* \frac{d^2\bar{r}(\alpha^1, \alpha^2, t)}{dt^2} = -\frac{\partial \bar{R}_1}{\partial \alpha^1}d\alpha^1 - \frac{\partial \bar{R}_2}{\partial \alpha^2}d\alpha^2 + \bar{p} + \bar{f}\rho_*, \tag{3.11}$$

where $d^2\bar{r}/dt^2$ is acceleration. Substituting $\bar{R}_i$ and ρ_* given by equations (3.3), (3.5) into (3.6), yields

$$\begin{aligned}
\rho\sqrt{a}\frac{d^2\bar{r}}{dt^2} &= \frac{\partial}{\partial \alpha^1}\left[(T^{11}\bar{e}_1 + T^{12}\bar{e}_2)\sqrt{a_{22}^*}\right]\\
&+ \frac{\partial}{\partial \alpha}\left[(T^{21}\bar{e}_1 + T^{22}\bar{e}_2)\sqrt{a_{11}^*}\right] + \bar{p}\sqrt{a^*} + \bar{f}\rho\sqrt{a}.
\end{aligned} \tag{3.12}$$

Let $\bar{G}_i$, $\bar{M}_p$ and $\bar{M}_f$ be the resultant moment vectors acting on the element of the shell defined by

$$\begin{aligned}
&\bar{G}_1 = \bar{r} \times \bar{R}_1, \qquad \bar{G}_2 = \bar{r} \times \bar{R}_2\\
&-\left(\bar{G}_1 + \frac{\partial \bar{G}_1}{\partial \alpha^1}d\alpha^1\right), \qquad -\left(\bar{G}_2 + \frac{\partial \bar{G}_2}{\partial \alpha^2}d\alpha^2\right) \qquad (\alpha^i + d\alpha^i = \text{const})\\
&\bar{M}_p = (\bar{r} \times \bar{p})\sqrt{a^*}\,d\alpha^1 d\alpha^2, \qquad \bar{M}_f = (\bar{r} \times \bar{f})\rho\sqrt{a^*}\,d\alpha^1 d\alpha^2.
\end{aligned} \tag{3.13}$$

Assuming the shell is in equilibrium, the sum of the moments vanishes. Hence

$$-\frac{\partial \bar{G}_1}{\partial \alpha^1}d\alpha^1 - \frac{\partial \bar{G}_2}{\partial \alpha^2}d\alpha^2 + \bar{M}_p + \bar{M}_q = 0. \tag{3.14}$$

Substituting $\bar{G}_i$, $\bar{M}_p$ and $\bar{M}_f$ in (3.13), we obtain

$$\left[-\left(\bar{r}\times\frac{\partial\bar{R}_1}{\partial\alpha}d\alpha^1\right)-\left(\bar{r}\times\frac{\partial\bar{R}_2}{\partial\alpha^2}d\alpha^2\right)+(\bar{r}\times\bar{p})+(\bar{r}\times\bar{f}\,)\right]$$
$$-(\bar{r}\times\bar{R}_1)d\alpha^1-(\bar{r}\times\bar{R}_2)d\alpha^2=0.$$

Further, using equation (3.5), we find

$$\begin{aligned}&[\bar{e}_1\times(T^{11}\bar{e}_1+T^{12}\bar{e}_2)+\bar{e}_2\times(T^{21}\bar{e}_1+T^{22}\bar{e}_2)]\sqrt{\overset{*}{a}_{11}}\sqrt{\overset{*}{a}_{22}}\,d\alpha^1 d\alpha^2\\&-\bar{r}\times\left[-\frac{\partial\bar{R}_1}{\partial\alpha^1}d\alpha^1-\frac{\partial\bar{R}_2}{\partial\underline{\alpha^2}}d\alpha^2+\bar{p}+\bar{f}\,\right]=0.\end{aligned} \tag{3.15}$$

Since the underlined term equals zero, we have

$$(\bar{e}_1\times\bar{e}_2)T^{12}+(\bar{e}_2\times\bar{e}_1)T^{21}=0. \tag{3.16}$$

It follows from the above that $T^{12}=T^{21}$. (Equations of motion of the soft biological shell in orthogonal Cartesian and cylindrical coordinates are given in appendix A.)

Remarks.

1. If a soft shell is parameterised along the principal axes then $T^{11}=T_1$, $T^{22}=T_2$, $T^{12}=0$, $\overset{*}{a}_{12}=a_{12}=0$ and the equation of motion (3.11) takes the form

$$\rho\sqrt{a}\frac{d^2\bar{r}}{dt^2}=\frac{\partial}{\partial\alpha^1}\left[T_1\sqrt{\overset{*}{a}_{22}}\,\bar{e}_1\right]+\frac{\partial}{\partial\alpha^2}\left[T_2\sqrt{\overset{*}{a}_{11}}\,\bar{e}_2\right]+\bar{p}\sqrt{a^*}+\bar{f}\rho\sqrt{a}. \tag{3.17}$$

2. During the dynamic process of deformation, different parts of the soft shell may undergo different stress–strain states. The biaxial stress–state occurs when $I^{(\mathbf{T})}{}_1 := T_1+T_2>0$, $I^{(\mathbf{T})}{}_2 := T_1T_2>0$, the uniaxial state develops in areas where $I^{(\mathbf{T})}{}_1>0$, $I^{(\mathbf{T})}{}_2=0$, and the zero stress–state takes place anywhere in the shell where $I^{(\mathbf{T})}{}_1=I^{(\mathbf{T})}{}_2=0$. The uniaxially stressed area ($T_2=0 \vee T_1=0$) will develop wrinkles oriented along the action of the positive principal membrane force $T_1 \vee T_2$, respectively. The equation of motion for the wrinkled zone becomes

$$\rho\sqrt{a}\frac{d^2\bar{r}}{dt^2}=\frac{\partial}{\partial\alpha_{1(2)}}\left[T_{1(2)}\sqrt{\overset{*}{a}_{22(11)}}\,\bar{e}_1\right]+\bar{p}\sqrt{a^*}+\bar{f}\rho\sqrt{a}. \tag{3.18}$$

To preserve the required smoothness and continuity of (the) surface S_*, the uniaxially stressed area is substituted by an 'ironed' surface made out of an array of closely packed reinforced fibres. Such an approach allows the equations of motion (3.11) to be used throughout the deformed surface.

3.2.3 Electromyogenic phenomenon

The propagation of electrical waves of depolarisation plays a key role in the initiation and coordination of mechanical processes in the bladder (–) i.e. the production of forces of contraction and relaxation. The detrusor has properties of the electrically anisotropic syncytium. Its electrical activity represents the integrated function of transmembrane voltage-dependent Ca^{2+} channels of L- and T-types, large Ca^{2+}-activated K^+ channels (BK_{Ca}), potential sensitive K^+ channels, and Cl^- ion channels. Although definite pacemaker cells have not been found in the UB, there are groups of cells responsible for generating electrical signals, V_p. The dynamics of their activity is described by

$$\alpha C_m^p \frac{dV_p}{dt} = -\sum_j I_j + I_{ext(i)}, \tag{3.19}$$

where C_m^p is the specific smooth muscle cell membrane capacitance, I_j are respective ion currents (j = L- and T-type Ca^{2+}, BK_{Ca}, voltage-dependent Na^+, K^+ and Cl^-), $I_{ext(i)}$ is the external membrane current. The ion currents are

$$\begin{aligned} &I_{Ca} = \frac{g_{Ca(i)} z_{Ca}\left(V_p - V_{Ca}\right)}{1 + \vartheta_{Ca}[Ca_i^{2+}]_p}, \; I_{Ca-K} = \frac{g_{Ca-K(i)}\, \rho_\infty\left(V_p - V_{Ca-K}\right)}{0.5 + [Ca_i^{2+}]_p} \\ &I_{Na} = g_{Na(i)} m_{Na}^3 h_{Na}\left(V_p - V_{Na}\right), \; I_K = g_{K(i)} n_K^4\left(V_p - V_K\right) \\ &I_{Cl} = g_{Cl(i)}\left(V_p - V_{Cl}\right), \end{aligned} \tag{3.20}$$

where V_{Ca}, V_{Ca-K}, V_{Na}, V_K, V_{Cl} are the reversal potentials, $g_{Ca(i)}$, $g_{Ca-K(i)}$, $g_{Na(i)}$, $g_{K(i)}$, $g_{Cl(i)}$ are the maximal conductances of the respective ion currents, ϑ_{Ca} is the numerical parameter, and the turnover of free intracellular calcium, Ca_i^{2+}, in the pacemaker cell yields

$$\frac{d[Ca_i^{2+}]_p}{dt} = \frac{0.2 z_{Ca}\left(V_p - V_{Ca}\right)}{1 + \vartheta_{Ca}[Ca_i^{2+}]_p} - 0.3[Ca_i^{2+}]_p. \tag{3.21}$$

The dynamic variables of the ion channels z_{Ca}, ρ_∞, m_{Na}, h_{Na}, n_K are given by

$$\begin{aligned} &dz_{Ca}/dt = (z_\infty - z_{Ca})/\tau_z, \quad dh_{Na}/dt = \lambda_h(h_\infty - h_{Na})/\tau_h \\ &dn_K/dt = \lambda_n(n_\infty - n_K)/\tau_n, \; \rho_\infty = \left[1 + \exp 0.15(V_p + 47)\right]^{-1} m_{Na} \\ &= m_\infty(V_p). \end{aligned} \tag{3.22}$$

In the above m_∞, h_∞, n_∞, z_∞ are calculated as

$$\begin{aligned} &y_\infty = \frac{\alpha_{y\infty}}{\alpha_{y\infty} + \beta_{y\infty}}, \; (y = m, h, n) \\ &z_\infty = [1 + \exp(-0.15(V + 42))]^{-1}, \end{aligned} \tag{3.23}$$

where

$$\alpha_{m\infty} = \frac{0.12\left(V_p + 27\right)}{1 - \exp\left(-\left(V_p + 27\right)/8\right)}, \quad \beta_{m\infty} = 4\exp\left(-\left(V_p + 47\right)/15\right)$$

$$\alpha_{h\infty} = 0.07\exp\left(-\left(V_p + 47\right)/17\right), \quad \beta_{h\infty} = \left[1 + \exp\left(-\left(V_p + 22\right)/8\right)\right]^{-1}$$

$$\alpha_{n\infty} = \frac{0.012\left(V_p + 12\right)}{1 - \exp\left(-\left(V_p + 12\right)/8\right)}, \quad \beta_{n\infty} = 0.125\exp\left(-\left(V_p + 20\right)/67\right).$$

The propagation of the wave of depolarisation, V_d, within the syncytium is described by

$$C_m^d \frac{\partial V_d}{\partial t} = -(I_{m1} + I_{m2} + I_{ion}), \tag{3.24}$$

where C_m^d is the membrane capacitance of the detrusor, I_{m1}, I_{m2}, I_{ion} are the transmembrane and total ion currents, respectively, given by

$$\begin{aligned} I_{m1} = & \frac{\tilde{\mu}_1 - \tilde{\mu}_2}{2\pi G(1 + \tilde{\mu}_1)(1 + \tilde{\mu}_2)} \iint \left(\hat{g}_{o1} \frac{\partial^2 V_d}{\partial X'^2} + \hat{g}_{o2} \frac{\partial^2 V_d}{\partial Y'^2} \right) \\ & \times \left[\frac{(\alpha_1 - \alpha_1')^2/G_1 - (\alpha_2 - \alpha_2')^2/G_2}{(\alpha_1 - \alpha_1')^2/G_1 + (\alpha_2 - \alpha_2')^2/G_2} \right] d\alpha_1' d\alpha_2', \end{aligned} \tag{3.25}$$

$$I_{m2} = \left(\hat{g}_{o1} \frac{\partial^2 V_d}{\partial \alpha_1'^2} + \hat{g}_{o2} \frac{\partial^2 V_d}{\partial \alpha_2'^2} \right) \left(\frac{\hat{g}_{i2}}{G_2} + \frac{2(\tilde{\mu}_1 - \tilde{\mu}_2)}{\pi(1 + \tilde{\mu}_1)(1 + \tilde{\mu}_2)} \tan^{-1} \sqrt{\frac{G_1}{G_2}} \right), \tag{3.26}$$

$$I_{ion} = \tilde{g}_{Na} \hat{m}^3 \hat{h}(V_d - \tilde{V}_{Na}) + \tilde{g}_K \hat{n}^4 (V_d - \tilde{V}_K) + \tilde{g}_{Cl}(V_d - \tilde{V}_{Cl}), \tag{3.27}$$

where

$$G_1 = \hat{g}_{i1} + \hat{g}_{o1}, \quad G_2 = \hat{g}_{i2} + \hat{g}_{o2}, \quad G = \sqrt{G_1 G_2},$$
$$\tilde{\mu}_1 = \hat{g}_{o1}/\hat{g}_{i1}, \quad \tilde{\mu}_2 = \hat{g}_{o2}/\hat{g}_{i2}.$$

In the above $\hat{g}^*_{i(0)}$, $\tilde{g}_{Na}$, $\tilde{g}_K$, $\tilde{g}_{Cl}$—maximal intracellular (subscript *i*), interstitial (subscript *0*) space and respective ion conductivities, $\tilde{V}_{Na}$, $\tilde{V}_K$, $\tilde{V}_{Cl}$—reversal potentials of Na^+, K^+, and Cl^- currents. The dynamics of the probability variables, the activation and deactivation parameters of the ion gates are given by

$$\begin{aligned} \frac{d\hat{m}}{dt} &= \hat{a}_m(1 - \hat{m}) - \beta_m \hat{m} \\ \frac{d\hat{h}}{dt} &= \hat{a}_h(1 - \hat{h}) - \beta_h \hat{h} \\ \frac{d\hat{n}}{dt} &= \hat{a}_n(1 - \hat{n}) - \beta_n \hat{n} \end{aligned} \tag{3.28}$$

with the activation $\hat{\alpha}_y$ and deactivation $\hat{\beta}_y$ ($y = \hat{m}, \hat{h}, \hat{n}$) parameters given by

$$\begin{aligned}
\hat{\alpha}_m &= \frac{0.005(V_d - V_{Na})}{\exp 0.01(V_d - V_{Na}) - 1}, \quad \hat{\beta}_m = 0.2\exp((V_d + V_{Na})/38) \\
\hat{\alpha}_h &= 0.014\exp\frac{-(V_d + V_{Na})}{20}, \quad \hat{\beta}_h = \frac{0.2}{\exp 0.2(V_{Na} - V_d) + 1} \\
\hat{\alpha}_n &= \frac{0.006(V_d - V_K)}{\exp 0.1(V_d - V_K) - 1}, \quad \hat{\beta}_m = 0.75\exp((V_K - V_d)/10).
\end{aligned} \tag{3.29}$$

Substituting equations (3.20) and (3.21) into (3.24), we obtain

$$\begin{aligned}
C_m^d\frac{\partial V_d}{\partial t} &= -\frac{\tilde{\mu}_1 - \tilde{\mu}_2}{2\pi G(1 + \tilde{\mu}_1)(1 + \tilde{\mu}_2)}\iint\left(\hat{g}_{o1}\frac{\partial^2 V_d}{\partial X'^2} + \hat{g}_{o2}\frac{\partial^2 V_d}{\partial Y'^2}\right) \\
&\times\left[\frac{(\alpha_1 - \alpha_1')^2/G_1 - (\alpha_2 - \alpha_2')^2/G_2}{(\alpha_1 - \alpha_1')^2/G_1 + (\alpha_2 - \alpha_2')^2/G_2}\right]d\alpha_1' d\alpha_2' \\
&-\left(\hat{g}_{o1}\frac{\partial^2 V_d}{\partial \alpha_1'^2} + \hat{g}_{o2}\frac{\partial^2 V_d}{\partial \alpha_2'^2}\right)\left(\frac{\hat{g}_{i2}}{G_2} + \frac{2(\tilde{\mu}_1 - \tilde{\mu}_2)}{\pi(1 + \tilde{\mu}_1)(1 + \tilde{\mu}_2)}\tan^{-1}\sqrt{\frac{G_1}{G_2}}\right) - I_{ion}.
\end{aligned} \tag{3.30}$$

The system of equations for the self-oscillatory electrical activity of the detrusor, V, is

$$ƛC_m\frac{dV}{dt} = -\sum_j \tilde{I}_j \tag{3.31}$$

where, $ƛ$ is the numerical parameter, C_m is the smooth muscle cell membrane capacitance, and $\tilde{I}_j$ is the sum of the respective ion currents

$$\begin{aligned}
&\tilde{I}_{Ca}^f = g_{Ca}^f \tilde{m}_I^3 \tilde{h}(V - V_{Ca}), \quad \tilde{I}_{Ca}^s = g_{Ca}^s \tilde{x}_{Ca}(V - V_{Ca}) \\
&\tilde{I}_{K1} = g_{K1}\tilde{n}^4(V - V_{K1}), \quad \tilde{I}_{Ca-K} = \frac{g_{Ca-K}^f[Ca_i^{2+}](V - V_{Ca})}{0.5 + [Ca_i^{2+}]} \\
&\tilde{I}_{Cl} = g_{Cl}(V - V_{Cl}).
\end{aligned} \tag{3.32}$$

Here V_{Ca}, V_{K1}, V_{Cl} are the reversal potentials, and $g_{Ca}^f, g_{Ca}^s, ..g_{Ca-K}, g_{Cl}$ are the maximal conductances for the ion currents, $\tilde{m}$, $\tilde{h}$, $\tilde{n}$ and $\tilde{x}_{Ca}$ are dynamic variables described by

$$\begin{aligned}
&m_I = \frac{\tilde{a}_m}{\tilde{a}_m + \tilde{\beta}_m}, \quad ƛh\frac{d\tilde{h}}{dt} = \tilde{\alpha}_h(1 - \tilde{h}) - \tilde{\beta}_h\tilde{h} \\
&ƛh\frac{d\tilde{n}}{dt} = \tilde{\alpha}_n(1 - \tilde{n}) - \tilde{\beta}_n\tilde{n}, \quad ƛ\tau_{xCa}\frac{d\tilde{x}_{Ca}}{dt} = \frac{1}{\exp(-0.15(V + 50))} - \tilde{x}_{Ca} \\
&ƛ\frac{d[Ca_i^{2+}]}{dt} = \wp_{Ca}\tilde{x}_{Ca}(V_{Ca} - V) - [Ca_i^{2+}],
\end{aligned} \tag{3.33}$$

where the activation $\tilde{\alpha}_y$ and deactivation $\tilde{\beta}_y(y = \tilde{m}, \tilde{h}, \tilde{n})$ parameters of ion channels satisfy the empirical relations

$$\begin{aligned}\tilde{\alpha}_{\tilde{m}} &= \frac{0.1(50 - \tilde{V})}{\exp(5\text{–}0.1\tilde{V}) - 1}, \ \tilde{\beta}_{\tilde{m}} = 4\exp((25 - \tilde{V})/18) \\ \tilde{\alpha}_{\tilde{h}} &= 0.07\exp((25\text{–}0.1\tilde{V})/20), \ \tilde{\beta}_{\tilde{h}} = [1 + \exp(5.5\text{–}0.1\tilde{V})]^{-1} \\ \tilde{\alpha}_{\tilde{n}} &= \frac{0.01(55 - \tilde{V})}{\exp(5.5\text{–}0.1\tilde{V}) - 1}, \ \tilde{\beta}_{\tilde{n}} = 0.125\exp((45 - \tilde{V})/80).\end{aligned} \tag{3.34}$$

Here $\tilde{V} = (127V + 8265)/105$, $\wp_{Ca}$ is the parameter referring to the dynamics of Ca^{2+} channels, τ_{xCa} is the time constant, $\hbar$ is a numerical constant.

The evolution of L- and T-type Ca^{2+}– channels depends on the wave of depolarisation, V_d, and is defined by

$$\begin{aligned} g_{Ca}^{s}(t) &= [\delta(V) + (\lambda(t) - 1)](\max g_{Ca}^{s}) \\ g_{Ca}^{f}(t) &= (\lambda(t) - 1)g_{Ca}^{f}, \end{aligned} \tag{3.35}$$

where

$$\lambda(t) \geqslant 1.0, \qquad \delta(V) = \begin{cases} 1, & \text{for} \quad V \geqslant V_d^* \\ 0, & \text{otherwise} \end{cases}.$$

Here V_d^* is the threshold value for V_d and λ is the stretch ratio.

3.2.4 Electromechanical coupling

Constitutive relations for the uniaxially stretched bioshell are functions of $\lambda_1 \vee \lambda_2$, empirical mechanical constants $c_i(i = \overline{1, m}$ and biological factors, Z_{ij}, which include the turnover of intracellular calcium ions, depolarisation of the cell membrane, etc.

$$\begin{aligned} T_1 &= f_1(\lambda_1, c_1, \ldots c_m, Z_{ij}) \text{ for } \lambda_1 > 1, \ \lambda_2 < 1 \\ T_2 &= f_2(\lambda_2, c_1, \ldots c_m, Z_{ij}) \text{ for } \lambda_1 < 1, \ \lambda_2 > 1. \end{aligned} \tag{3.36}$$

In the in-plane biaxial state, $\lambda_1 > 1$ & $\lambda_2 > 1$ ($T_1 > 0$ & $T_2 > 0$) relations (3.36) have the form

$$\begin{aligned} T_1 &= F_1(\lambda_1, \lambda_2, c_1, \ldots c_m, Z_{ij}) \\ T_2 &= F_2(\lambda_1, \lambda_2, c_1, \ldots c_m, Z_{ij}). \end{aligned} \tag{3.37}$$

(Note: in general $f_n(\ldots) \neq F_n(\ldots)$, however, $f_n(\ldots)$ can be defined uniquely if $F_n(\ldots)$ is known). These can be decomposed as the sum of the 'passive', $T_{c,\,l}^{p}(\lambda_1, Z_2, c_i)$, and the 'active', $T_{c,\,l}^{a}(\lambda_{1/2}, Z_{ij}, c_i)$, components

$$T_{1/2} = T_{c,1}^{p}(\lambda_1, \lambda_2, c_i) + T_{c,1}^{a}(\lambda_{1/2}, Z_{ij}, c_i). \tag{3.38}$$

The 'passive' component reflects the mechanical characteristics of inactive SM cells, collagen and elastin fibres, whilst the 'active' accounts for intracellular contractile protein mechanics. Although exact data on the sequence of mechanical events in the muscle syncytia during contractions are not available, it is reasonable to assume that both muscle syncytia contract simultaneously. Such coordination results in the generation of strong active forces $T^{a}_{c,\,l}$ needed to effectively expel urine from the bladder. This assumption, working with the fine fibrillar structure of smooth muscle, suggests that $T^{a}_{c,\,l}$ are produced only in the preferred directions, either longitudinal or circumferential, and as such can be characterised in full by uniaxial tests.

Based on the fact that the key player in the transformation of microscopic properties of electrical excitatory events into macroscopic contractions is Ca_i^{2+}, as a first approximation of the excitation–contraction coupling, the active force—intracellular Ca^{2+} relationship—can be suggested in the experimentally supported form

$$T^{a}_{c,l} = \begin{cases} 0, & [Ca_i^{2+}] \leq 0.1\mu M \\ c_7 + c_8[Ca_i^{2+}]^4 + c_9[Ca_i^{2+}]^3 + c_{10}[Ca_i^{2+}]^2 + c_{11}[Ca_i^{2+}], & 0.1 < [Ca_i^{2+}] \leq 1\mu M \\ \max T^a, & [Ca_i^{2+}] > 1\,\mu M \end{cases}. \tag{3.39}$$

3.2.5 Mechanosensory element

Information about the stress–strain state of the bioshell is detected and processed by mechanoreceptors. Embedded in the bladder wall, these respond to stretching by the generation of nerve ending, V_{se}, and dendritic, V_{sd}, potentials

$$\begin{aligned} C_{se}\frac{dV_{se}}{dt} &= -(I^s_{Na} + I^s_K + I^s_{Cl}) - (V_{sd} - V_{se})/R_{se} \\ \gamma\frac{dV_{sd}}{dt} &= -V_{sd} + k(V_{sd} - V_{se})/R_{se}, \end{aligned} \tag{3.40}$$

where C_{se} is the dendritic membrane capacitance, R_{se} is the membrane resistance and γ, k are the membrane time and numerical constants. I^s_{Na}, I^s_K, I^s_{Cl} are the ion currents given by

$$\begin{aligned} I^s_{Na} &= g^s_{Na} m_s^3 h_s (V_{se} - V^s_{Na}), \quad I^s_K = g^s_K n_s^4 (V_{se} - V^s_K) \\ I^s_{Cl} &= g^s_{Cl}(V_{se} - V_{Cl}). \end{aligned} \tag{3.41}$$

Here g^s_{Na}, g^s_K, g^s_{Cl} are the maximal conductances for Na^+, K^+, Cl^- channels, m_s, h_s, n_s are the probabilities of opening these channels and V^s_{Na}, V^s_K, V^s_{Cl} are the reversal potentials of the respective ion currents. The activation, α_{y_*}, and deactivation, β_{y_*}, of ion channels is described by

$$\frac{dy^*}{dt} = \psi(\alpha_{y*}(1 - y^*) - \beta_{y*}y^*) \qquad (y^* = m_s, h_s, n_s). \tag{3.42}$$

Here

$$\alpha_{ms} = 0.221\exp((\lambda(t) - 1) + 0.01V_{se}),\ \beta_{ms} = 4\exp(-V_{se}/18)$$
$$\alpha_{hs} = 0.014\exp(-V_{se}/20),\ \beta_{hs} = 0.2/(1 + \exp(3-0.1V_{se}))\tag{3.43}$$
$$\alpha_{ns} = 0.1\exp(1-0.1V_{se}),\ \beta_{ns} = 0.185\exp(-V_{se}/80)$$

and ψ is the temperature scale factor.

The generated dendritic action potential triggers the propagation of a wave of depolarisation along the myelinated fibre to the secondary-sensory neuron in the spinal cord.

3.2.6 Problem closure

Anatomical and physiological observations serve as the basics to formulate the initial and boundary conditions. For example, assuming that: (i) the initial undeformed configuration of the urinary bladder is defined by the intraluminal pressure, p, (ii) the bladder sphincter remains rigidly fixed throughout filling and voiding, (iii) the organ is at electromyogenic rest, (iv) the *a priori* defined pacemaker cell discharges action potentials of given intensity and duration, then the mathematical model as stated above can accurately reproduce the propagation of electrical waves of depolarisation and induced mechanical waves of contraction-relaxation in the bioshell and predict changes in the configuration of the organ during the dynamic response.

References

Argungu M *et al* 2018 Modelling afferent nerve responses to bladder filling arXiv:1510.03119

Bastianssen E H C, van Leeuwen J L, Vanderschoot J and Redert P A 1996b A myocybernetic model of the lower urinary tract *J. Theor. Biol.* **178** 113–33

Bastianssen E H C, Vanderschoot J and van Leeuwen J L 1996a State–space analysis of a myocybernetic model of the lower urinary tract *J. Theor. Biol.* **180** 215–27

Celik I B, Varol A, Bayrak C and Nanduri J R 2007 A one dimensional mathematical model for urodynamics *Proc. FEDSM2007, 5th ASME/JSME Fluids Eng. Conf. (San Diego, CA, 30 July–2 August 2007)* pp 1–7

Damaser M S 1999 Whole bladder mechanics during filling *Scand. J. Urol. Nephrol. Suppl.* **201** 51–8

Damaser M S and Lehman S L 1993 Does it matter, the shape of the bladder? *Neurourol. Urodynam.* **12** 227–80

De Pascalis R, Abrahams I D and Parnell W J 2014 On nonlinear viscoelastic deformations: a reappraisal of Fung's quasi-linear viscoelastic model *Proc. Royal. Soc.* A **470** 20140058

Fernández D R, Chamizo J M G, Pérez F M and Payá A S 2004 Modeling the distributed control of the lower urinary tract using a multiagent system *Modeling Decisions for Artificial Intelligence* (*Lecture Notes in Comp. Sci.* vol 3131) (Berlin: Springer) pp 1–5

Fletcher W, Smith F T and Fry C 1997 A computer simulation of micturition *Private report*

Fry C H, Sadananda P, Wood D N, Thiruchelvam N, Jabr R I and Clayton R 2011 Modeling the urinary tract - computational, physical, and biological methods *Neurourol. Urodynam.* **309** 692–9

Hosein R A and Griffiths D J 1990 Computer simulation of the neural control of bladder and urethra *Neurourol. Urodynam.* **9** 601–18

Kim J, Lee M K and Choi B 2011 A study of the fluid mechanical urinary bladder simulator and reproduction on human urodynamics *Int. J. Precision. Eng. Manufact.* **12** 679–85

Kinder M V, Bastiaanssen E H C, Janknegt R A and Marani E 1999 The neuronal control of the lower urinary tract: a model of architecture and control mechanisms *Arch. Physiol. Biochem.* **107** 203–22

Korkmaz I and Rogg B 2007 A simple fluid-mechanical model for prediction of the stress-strain relation of the male urinary bladder *J. Biomech.* **40** 663–8

Křen J, Horák M, Zát'ura F and Rosenberg M 2001 Mathematical model of the male urinary tract *Biomed. Pap.* **145** 91–6

McKean H P 1970 Nagumo's equation *Adv. Math.* **4** 209–23

Miftahof R N and Akhmadeev N R 2019 *Soft Biological Shells in Bioengineering* (Bristol: IOP Publishing)

Pérez F M, Chamizo J M G, Payá A S and Fernández D R 2008 A robust model of the neuronal regulator of the lower urinary tract based on artificial neural networks *Neurocomputing* **71** 743–54

Regnier C, Kolsky H, Richardson P D, Ghoinem G M and Susset J G 1983 The elastic behavior of the urinary bladder for large deformations *J. Biomech.* **16** 915–22

Spirka T A and Damaser M S 2007 Modelling of physiology of the urinary tract *J. Endourol.* **21** 294–9

van Beek A J 1997 A finite element model of the urinary bladder *Priv. Rep.*

van Duin F, Rosier P F W M, Benelmans B L H, Wijkstra H, Debruyne F M J and van Oosterom A 2000 Comparison of different computer models of the neural control system of the lower urinary tract *Neurourol. Neurodynam.* **12** 289–310

Vlastelica I, Veljkovic D, Stojanovic B, Rosic M and Kojic M 2007 Modeling urinary bladder deformation within passive and active regimes *J. Serb. Soc. Comput. Mech.* **1** 129–34

Zang X J, Li X Y and Wang J L 2010 Computational fluid dynamics model of bladder-urethra system for SUI *IFMBE Proc (6th World Cong Biomech)* **31** 1495–8

The Brain–Bladder Axis in Tissue Growth and Remodelling

Roustem N Miftahof and Christian J Cyron

Chapter 4

Signalling mechanisms

The single biggest problem in communication is the illusion that it has taken place.

— George B Shaw

Neurohormonal signalling in bladder tissue growth and remodelling is analysed with special attention placed on neurohormonal transmission by L-glutamate, GABA, glycine, SS, GHRH, Ghr, GH and IFG-1. The effects of their co-localisation and co-release and the polymodality of their respective receptors on myogenesis and collagenesis in the bladder are modelled.

4.1 L-Glutamate

The modes of communication of physiological and pathological stimuli among the brain–urinary bladder cytoelements are neuro- and endocrine transmission. Although there is a plethora of neurotransmitters and hormones that directly and indirectly contribute to the transfer of signals, our focus will fall on those essential for tissue growth and remodelling in the wall of the urinary bladder.

Amino acid L-glutamate is the main excitatory neurotransmitter in the brain. Although stored in and released from neurons, its biosynthesis depends on astrocytic production and delivery of the precursor glutamine. The intracellular source of conversion of glutamine to Glu by glutamine synthetase is: (i) Glu cleared from the synaptic cleft and translocated into the cell via the excitatory amino acid (EAAT) and the high-affinity membrane Glu transporters (Danbolt 2001, Shigeri *et al* 2004), (ii) Glu 'recycled' from re-uptaken extracellular γ-aminobutyric acid via GABA transporter protein 3 (GAT_3), and (iii) Glu synthesised *de novo* in astrocytes from glucose and amino acid derivatives. Astrocytic glutamine is actively moved into

doi:10.1088/978-0-7503-3567-6ch4

neurons via the unidirectional system-A transporter where after the reaction with phosphate-activated glutaminase, Glu is finally formed.

Cytosolic Glu is packaged and stored in synaptic vesicles (SVs) of the nerve terminals. This is provided by $VGluT_{1/2}$ and the electrochemical force of a H^+-ATPase generated proton gradient (Blakely and Edwards 2012). The estimated fill rate time constant for Glu in intact synapses is ~7 s (Hori and Takahashi 2012). There are three mechanisms responsible for the release of the neurotransmitter from SVs into the synaptic cleft. The prime one is exocytosis in response to the depolarisation of the presynaptic membrane. This involves a complex sequence of the positioning, docking and formation of fusion-opening pore processes. These are tightly operated by *N*-ethylmaleimide-sensitive factor attachment protein receptors, Sec1/Munc 18-like protein, and Ca^{2+}-dependent activator protein for secretion and synaptotagmin (Savigny *et al* 2007, Pang and Südhof 2010, Südhof 2013, Kreutzberger *et al* 2017). It has been estimated that ~100–250 μM is released (Clemens *et al* 1992). The remnants of SVs either merge with or close rapidly and undock from the synaptic membrane (Hamper *et al* 2018). The second obligate mechanism of non-vesicular Glu release that also maintains intracellular physiological levels of Glu and defends the cell against oxidative stress, is a Na^+-independent cystine/glutamate antiporter system. It mediates the uptake of the anionic form of cystine into the cell in exchange for glutamate in a 1:1 ratio (Albrecht *et al* 2010). The third mechanism relies on the reverse action of the membrane-bound transporter A system.

A part of the released Glu reaches the postsynaptic membrane in <10 μs where it binds to respective receptors (GluRs) at the rate of 1 $(\mu M\ s)^{-1}$ (Clemens *et al* 1992, Diamond and Jahr 1997). The other part spreads throughout the synaptic cleft over 50–100 μs and is either enzymatically degraded by glutamate pyruvate transaminase (GPT), glutamine synthetase, and glutamate dehydrogenase (Matthews *et al* 2000), and/or is buffered, at a rate of 1–20 μs, and translocated into astrocytes and neurons by the active transporter, $EAAT_3$ (Diamond and Jahr 1997). The process is driven by the transmembrane electrochemical gradient with the suggested stoichiometry: an intracellular co-transport of Glu, a single H^+ proton and $3Na^+$ ions in exchange for one K^+ ion (Rowley *et al* 2012).

Glutamate receptors comprise eight members of G-protein coupled receptors (GPCR) (metabotropic) and receptor channels (ionotropic). The assembly of the latter is based on their agonist selectivity into four major types: N-methyl-D-aspartate (NMDA), α-amino-3-hydroxy-5-methyl-4-isoxazole-propionic acid (AMPA), kainate (KA) and δ receptor channels. Molecular profiling has revealed seven subunits for NMDA ($NMDAR_1/_{2A\text{-}2D/3A\text{-}B}$), four for AMPA ($Glu_{1\text{–}4}$) and five for KA receptors ($Glu_{5\text{–}7}$, $KA_{1/2}$). Although successfully cloned, the role of the δ receptors in the body is still unclear (Stojilkovic *et al* 2010, Pin and Acher 2002, Mayer and Armstrong 2004).

Depending on the arrangement of tetrameric complexes from individual/ heteromeric subunits, there is a great diversity of channel properties. Thus NMDA receptors, both ligand- and voltage-gated, exhibit a different excitation behaviour when compared to AMPA and KA receptors. Their full activation requires the obligatory binding of Glu and Gly to the reactive sites with concomittant allosteric modulation by Mg^{2+}, Zn^{2+} and H^+ of the channel opening and regulation of cations (Ca^{2+}, Na^+) influx (Traynelis *et al* 1995, Nowak *et al* 1984, Hamper *et al* 2018). AMPA receptors also form tetrameric complexes that mediate fast excitatory neurotransmission whereas homomeric $KA_{1/2}$ receptors are virtually inactive and serve a modulatory function only.

Metabotropic Glu receptors ($mGluR_{1-8}$) are divided, based on amino acid morphology, into three groups. Group 1 ($mGluR_{1/5}$) receptors are linked to excitatory $G_{\alpha s/\alpha 11}$ proteins and, after binding to postsynaptic Glu, elicit a positive effect on protein kinase C (PkC) with the release on Ca^{2+} from the endoplasmic reticulum via the PLC pathway. By contrast, groups 2 and 3 ($mGluR_{2/3}$ and $GluR_{4-8}$, respectively) are coupled to inhibitory $G_{\alpha i/\alpha o}$-proteins and after activation, exert a negative effect on protein kinase A (PkA) by affecting the cyclic adenosine monophosphate (cAMP) pathway (see appendix B for details). Although the precise mechanisms have not yet been fully elucidated, there is experimental evidence suggesting that mGluRs modulates presynaptic voltage-dependent Ca^{2+} channels on Glu and GABA-containing nerve terminals, thereby influencing the neurotranmitter release (Takahashi *et al* 1996).

After dissociation from the Glu-mGluR reactive complex, glutamate is enzymatically metabolised or taken up by astrocytes whilst receptors desensitise at a rate of ~5 ms and are further internalised.

Results of *in situ* hybridisation studies have shown strong gene expressions for: (i) $NMDAR_1$, $GluR_{1/2}$ type receptors in neurons in the SDH, anterior and intermediate regions in the PeVN (van den Pol *et al* 1994, Eyigor *et al* 2001); (ii) $GluR_1$ and $GluR_{2/3}$ receptors in the ArcN and the ME, respectively, $NMDA_1$ and $mGluR_{1a}$ receptors separately and together with cholinergic muscarinic μ_1 and $GABA_A$ receptors throughout the ArcN-ME circumventricular organ (Petralia and Wenthold 1992, van den Pol *et al* 1994, Kiss *et al* 1997, Kiss *et al* 2006, Baccam *et al* 2007, Zhang and van den Pol 2015); (iii) $NMDA_1$ receptor (a small fraction only) in somatotrophs (Bhat *et al* 1995).

The state diagram of neurotransmission by glutamate is shown in figure 4.1.

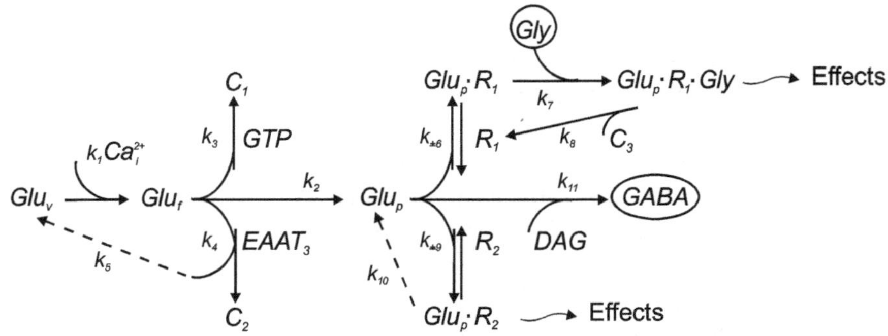

Figure 4.1. General scheme of glutamate neurotransmission. Here: $k_{\pm i}$ are the rate constants of forward (+) and backward (−) chemical reactions; $R_1 := NMDA \cdot R$, $R_2 := AMPA \cdot R$ and the meaning of other substrates is as described above. Ligands in circles are involved in cross-talk with other signalling pathways.

Assuming all reactions satisfy the Michaelis–Menten kinetics, the corresponding system of equations describing the dynamics of substrates interactions and conversions yields

$$d\mathbf{X}_{Gl}/dt = \mathbf{A}_l\mathbf{X}_{Gl}(t) \tag{4.1}$$

where $\mathbf{A}_l$ is the matrix of transfer rate constants and the vector $\mathbf{X}_{Gl}(t) = (x_{1-13})^{\mathrm{T}}$ is defined by

$$\begin{aligned}
&x_1 = [Glu_{\mathrm{v}}], \quad x_2 = [Glu_{\mathrm{f}}], \quad x_3 = [Glu_{\mathrm{p}}]\\
&x_4 = [GTP], \quad x_5 = [EAAT_3], \quad x_6 = [Glu_{\mathrm{f}} \cdot EAAT_3]\\
&x_7 = [NMDA \cdot R], \quad x_8 = [Glu_{\mathrm{p}} \cdot NMDA \cdot R], \quad x_9 = [Gly]\\
&x_{10} = [Glu_{\mathrm{p}} \cdot NMDA \cdot R \cdot Gly], \quad x_{11} = [AMPA \cdot R]\\
&x_{12} = [Glu_{\mathrm{p}} \cdot AMPA \cdot R], \quad x_{13} = [GAD].
\end{aligned}$$

The quantitative assessment of $k_{\pm i}$ shows that the system is a combination of rapid equilibrium and slow rate-limiting segments. Therefore, applying the state-stationary approximations to equation (4.1), it can be reduced to

$$d\mathbf{X}_{Gl}/dt = \mathbf{A}_l\mathbf{X}_{Gl}(t) + \mathbf{C}_{Gl}. \tag{4.2}$$

Here $\mathbf{X}_{Gl}(t) = (x_{1-4}, x_9, x_{13})^{\mathrm{T}}$, $\mathbf{C}_{Gl} = (k_5x_6, 0, k_{-6}x_8 + k_{-9}x_{12}, 0, 0, 0)^{\mathrm{T}}$ and the non-zero elements of the matrix $\mathbf{A}_l(6 \times 6)$ are

$$\begin{aligned}
&a_{11} = -a_{21} = -k_1g_{\mathrm{syn}}^{\mathrm{Ca}}V^{\mathrm{f}}(t)[\mathrm{Ca}_0^{2+}], \quad a_{22} = -(k_2 + k_3x_4 + k_4x_5)\\
&a_{32} = k_2, \quad a_{33} = -(k_6x_7 - x_8) + k_9x_{11} + k_{11}x_{13})\\
&a_{44} = -k_3x_2, \quad a_{55} = -k_7x_8, \quad a_{66} = -k_{11}x_3.
\end{aligned}$$

Concentrations of the substrates, x_{5-8}, x_{10-12}, in a quasi-steady-equilibrium state are calculated from the algebraic equations

$$\begin{aligned}
x_5(t) &= \frac{k_5x_5^0}{k_5 + k_4x_2}\\
x_6(t) &= \frac{k_4x_5^0x_2}{k_5 + k_4x_2}\\
x_7(t) &= \frac{k_8(k_{-6} + k_7x_9)x_7^0}{k_8(k_{-6} + k_7x_9) + k_6(k_8 + k_7x_9)x_3}\\
x_8(t) &= \frac{k_6k_8x_7^0x_3}{k_8(k_{-6} + k_7x_9) + k_6(k_8 + k_7x_9)x_3}\\
x_{10}(t) &= \frac{k_7x_8x_9}{k_8}\\
x_{11}(t) &= \frac{(k_{-9} + k_{10})x_{11}^0}{k_{-9} + k_9k_{10}x_3}\\
x_{12}(t) &= \frac{k_9x_{11}^0x_3}{k_{-9} + k_9k_{10}x_3}.
\end{aligned} \tag{4.3}$$

Hereinafter the total concentrations of substrates that remain constant in the system are denoted by: x_i^0. Assuming that the initial state of the synapse and the reacting components are known, $\mathbf{X}_{Gl}(0) = \mathbf{X}_{Gl0}$, equations (4.2) and (4.3) provide mathematical formulation of the dynamics of neurotransmission by glutamate.

4.2 GABA

γ-Aminobyturic acid is the principal inhibitory neurotranmsitter in the CNS. It is generated only in GABAergic cells that contain the glutamic acid decarboxylase (GAD) enzyme, these being mainly neurons and pancreatic islets. The immediate precursor for the neurotransmitter synthesis is Glu which can be produced by the reaction of: (i) converting astrocytic delivered glutamine by phosphate-activated glutaminase, or (ii) transamination of intracellularly produced α-KG by α-oxoglutarate transaminase. GABA is formed finally by the decarboxylation of Glu by GAD.

As with any other neurotransmitter, GABA is transported to the nerve terminal for storage in SVs. This process is maintained by the vesicular GABA transporter (VGAT) and energised by vacuolar H^+-ATPase. Using internal reflection fluorescent microscopy combined with a potentiometric assay and a neuron-specific expression of genetically encoded pH sensor strategy, it has been demonstrated that the sequesteration depends entirely on the GABA/H^+ antiport and VGAT functions as its exchanger by sustaining a higher state pH. This differs from Glu packaging where additional Cl^-/H^+ and K^+/H^+ exchange mechanisms are required and explains the slower fill rate time ~25–40 (s) for GABA when compared to Glu (Farsi *et al* 2016, Yamashita *et al* 2018, Herman *et al* 2018).

The release mechanisms of GABA from SVs are identical to Glu. The estimated time required varies between 80 and 120 (ms). Free GABA in the synaptic cleft is subjected to transmembrane GABA transporters ($GAT_{1/3}$). These move the neurotransmitter intracellularly by utilising the inward Na^+ gradient with the suggested stoichiometry 2 or $3Na^+$, $1Cl^-$ ions and one GABA molecule (Radian *et al* 1986, Kanner 1994, Roth and Draguhn 2012). $GAT_{1/3}$ transporters are widespread throughout the brain with a characteristic pattern of distribution. Thus GAT_1 is found mainly on presynaptic GABAergic neurons and in close proximity to axonal terminals whilst the prime location of GAT_3 is astrocytes and, to a lesser extent, neurons (Radian *et al* 1990, Durkin *et al* 1995, Minelli *et al* 1996, Scimemi 2014). Once inside the cell, GABA undergoes enzymatic degradation by a series of 4-aminobutyrate-transaminases, i.e., 2-oxoglutarate, pyruvate and glyoxylate, with the production of succinate semialdehyde and Glu, L-alanine and Gly, respectively.

Postsynaptic GABA binds to the receptors on adjacent cells. There are three structurally and pharmacologically distinct classes, i.e. ligand-gated (ionotropic) $GABA_{A/C}$ chloride channels and G-protein coupled (metabotropic) $GABA_B$ receptors. The $GABA_{A/C}$ receptors are pentameric and trimeric assemblies, respectively, of subunits α1–6, β1–3, γ1–3, δ, ε and π. The major type A receptor isoform consists of 2α, 1β and 2γ, and type C – β1–3 subunits (Farrant and Nusser 2005). Arranged around a central ion pore, they form a channel exclusively permeable to Cl^- ions

(Siegel and Steinmann 2012, Barberis *et al* 2007). The formation of the GABA-$GABA_{A/C}$ receptor complex causes rapid and transient influx of anions and hyperpolarization of the cell membrane (phasic inhibition) (Farrant and Nusser 2005, Labrakakis *et al* 2014, Dixon *et al* 2014).

Metabotropic $GABA_B$ receptors are heterodimers composed of two subunits. Coupled to $G_{\alpha i}$-proteins, these produce their effects via the inhibiting adenylyl cyclase pathway and by impeding the opening of voltage-gated Ca^{2+} channels. In addition, the intermediate product, $G_{\beta/\gamma}$-subunit, activates rectifying K^+ and hinders Ca^{2+} channels (Tedfort and Zamponi 2006).

All three GABA receptor classes are expressed in: (i) the PeVN, where GAD-positive vesicles in axo-somatic and axo-dendritic synapses on SS-labelled neurons have been detected (Kakucska *et al* 1988); (ii) the ArcN-ME complex, mainly in the nerve terminals (Hrabovsky *et al* 2005, Betley *et al* 2013, Campbell *et al* 2017); and (iii) somatotrophs of the adenohypophysis (Zemkova and Stojilkovic 2018).

The statements related to neurotransmission by GABA are shown in figure 4.2.

The chemical kinetic rate equations related to the above scheme are given by

$$d\mathbf{X}_G/dt = \mathbf{A}_2\mathbf{X}_G(t) \tag{4.4}$$

where $\mathbf{A}_2$ is the rate constant matrix and the vector $\mathbf{X}_G(t) = (x_{14-26})^{\mathrm{T}}$ is

$$\begin{aligned}
&x_{14} = [GABA_{\mathrm{v}}], \quad x_{15} = [GABA_{\mathrm{f}}], \quad x_{16} = [GABA_{\mathrm{p}}] \\
&x_{17} = [GAT_{1,2,3}], \quad x_{18} = [GABA_A \cdot R], \quad x_{19} = [GABA_{\mathrm{p}} \cdot GABA_A \cdot R] \\
&x_{20} = [sGC], \quad x_{21} = [sGC^{\mathrm{A}}], \quad x_{22} = [GTP] \\
&x_{23} = [cGMP], \quad x_{24} = [PkG], \quad x_{25} = [PkG^{\mathrm{A}}] \\
&x_{26} = [PDE].
\end{aligned}$$

Hereinafter the superscript x_i^{A} indicates the active (phosphorylated) form of the respective substrate (ligand). The system (4.4) can be simplified to

$$d\mathbf{X}_G/dt = \mathbf{A}_2\mathbf{X}_G(t) + \mathbf{C}_G. \tag{4.5}$$

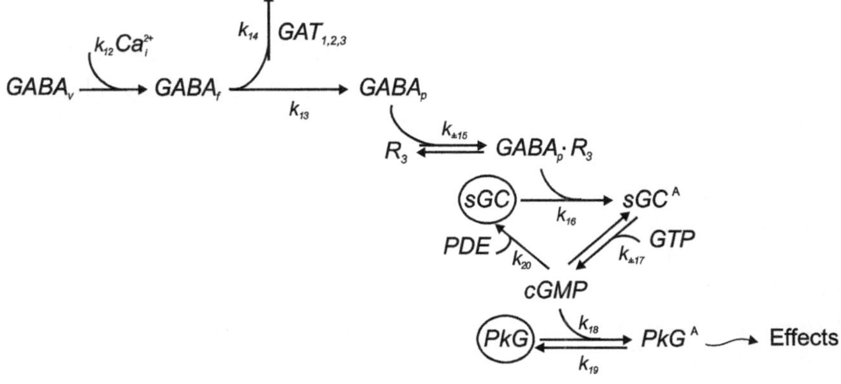

Figure 4.2. Scheme of neurotransmission by GABA. Here R_3 refers to the ionotropic $GABA_A$ receptor and the meaning of other substrates is described in the text.

Here the vector of unknowns is $\mathbf{X}_G(t) = (x_{14-17}, x_{22}, x_{23}, x_{26})^{\mathrm{T}}$ and the vector $\mathbf{C}_G = (C_{G(1)}, 0, C_{G(3)}, 0, C_{G(5)}, 0, 0)^{\mathrm{T}}$ has the components $C_{G(1)} = k_{11}x_3x_{13}$, $C_{G(3)} = k_{-15}x_{19}$, $C_{G(5)} = k_{-26}x_{36}$.

The non-zero elements of the reduced matrix $\mathbf{A}_2(7 \times 7)$ are

$$b_{11} = -b_{21} = -k_{12}g_{\mathrm{syn}}^{\mathrm{Ca}}V^{\mathrm{f}}(t)[\mathrm{Ca}_0^{2+}], \quad b_{22} = -k_{13}, \quad b_{24} = -b_{44} = k_{14}x_{15}$$
$$b_{32} = k_{13}, \quad b_{33} = -k_{15}x_{18}, \quad b_{55} = -(k_{17}x_{21} + k_{26}x_{34} + k_{46}x_{58})$$
$$b_{56} = k_{-17}x_{20}, \quad b_{65} = k_{17}x_{21}, \quad b_{66} = -(k_{-17}x_{20} + k_{18}x_{24} + k_{20}x_{26})$$
$$b_{77} = -(k_{20}x_{23} + k_{43}x_{52}).$$

The reactants $x_{18-21}, x_{24,25}$ are obtained from

$$\begin{aligned}
x_{18}(t) &= \frac{(k_{-15} + k_{16}x_{20})x_{18}^0}{k_{-15} + k_{15}x_{16} + k_{16}x_{20}} \\
x_{19}(t) &= \frac{k_{15}k_{16}x_{18}^0}{k_{-15} + k_{15}x_{16} + k_{16}x_{20}} \\
x_{20}(t) &= \frac{k_{17}x_{20}^0x_{22}}{k_{16}x_{19} + k_{-17}x_{23} + k_{17}x_{22}} \\
x_{21}(t) &= \frac{(k_{16}x_{19} + k_{-17}x_{23})x_{20}^0}{k_{16}x_{19} + k_{-17}x_{23} + k_{17}x_{22}} \\
x_{24}(t) &= \frac{(k_{19} + k_{50}x_{61})x_{24}^0}{k_{19} + k_{18}x_{23} + k_{50}x_{61}} \\
x_{25}(t) &= \frac{k_{18}x_{24}^0x_{23}}{k_{19} + k_{18}x_{23} + k_{50}x_{61}}.
\end{aligned} \tag{4.6}$$

Given the initial concentrations of the substrates, $\mathbf{X}_G(0) = \mathbf{X}_{G0}$, equations (4.5)–(4.6) describe the dynamics of the neurotransmission by GABA.

4.3 Glycine

Glycine is formed: (i) from serine by the enzyme glycine hydroxymethyl-transferase (GlyHt), and (ii) the reaction of transamination of GABA by glyoxulate. Its presence has been demonstrated in the brainstem and the dorsal horn of the spinal cord (lamina I and II) where it is co-localised and co-released together with GABA from a single vesicle (Aubrey and Supplisson 2018). As a neurotransmitter, Gly potentiates the excitatory effect of Glu by increasing the occupancy of the NMDA receptor coagonist site (Ahmadi *et al* 2003). Thus Gly serves dual inhibitory and excitatory functions. Its effects are terminated by re-uptake via high-affinity Gly transporters and enzymatic degradation by Gly-decarboxylase complex, D-amino acid oxydase (GlyDc) (figure 4.3).

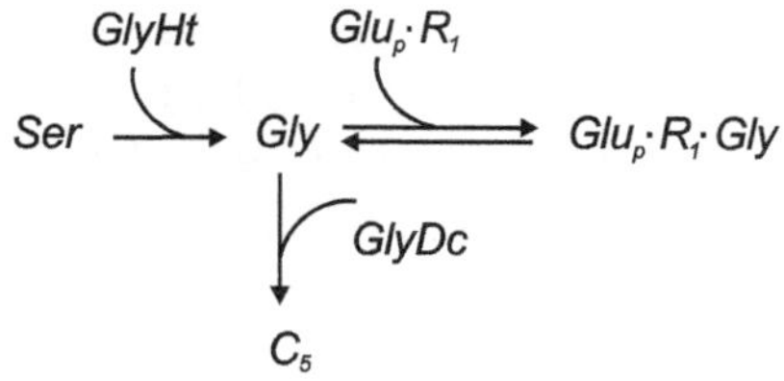

Figure 4.3. Scheme of glycine synthesis and degradation.

No Gly receptors have been identified in the PeVN, ArcN-ME circumventricular organ or the anterior pituitary gland. Therefore, for (the) modelling purposes, it will be assumed that Gly is available in physiological quantities in the system.

4.4 Somatostatin

Somatostatins, cyclic tetradecapeptides, are ubiquitous throughout the human body. Acting as neuro/autocrine/paracrine hormones and neurotransmitters, these influence both endocrine and exocrine secretion (Weckbecker *et al* 2003). In the brain the hormone is synthesised from a single precursor, the 116 amino acid preprohormone, predominantly in the anterior hypothalamus. After signal peptide removal, a single protein cleavage produces a bioactive peptide of 14 amino acids with a length and molecular weight of ~1.64 kDa (SS-14) (Liguz-Lecznar *et al* 2016, Martinez 2013). Interestingly, precursors of the neuropeptide have also been detected in the anterior pituitary gland, suggesting that it could be produced there as well (Levy *et al* 1993).

The hormone is stored in LDCVs, 80–110 nm in diameter, clustered in cell bodies, axons and dendrites (Barnett 2003). It is secreted into the hypothalamus with a 3.3 h periodicity having a pulsatile manner at a frequency of 3–4 cycles/h (Plotzky and Vale 1985, Tannenbaum and Ling 1984). The long-lasting high-frequency depolarisation, GH, IGF-1, GHRH, Ghr, adrenergic agonists, etc stimulate SS release, whilst cholinergic μ_1-receptor agonists can either facilitate or impede, with glucose, VIP, GABA and somatostatin itself inhibiting the neuropeptide exocytosis (Zigman *et al* 2006, Inenaga *et al* 1986, Eigler and Ben-Shlomo 2014, Chen *et al* 2015). Somatostatin is secreted extrasynaptically in a Ca^{2+}-dependent manner via volume transmission. The repetitive electrical stimulation causes a transient influx and endocytosis of Ca^{2+} ions from the endoplasmic reticulum. Once the concentration of free cytosolic Ca^{2+} reaches 50–100 $\mu M\ l^{-1}$, the movement of LDCVs starts *en masse*. These fuse at the cell surface membrane to release their content into the extracellular space (De-Miguel and Nicholls 2015). Owing to their size, molecules of SS cannot pass easily through the fusion pore thus slowing down the process, <50 ms, when compared to the transmission by Glu or GABA which occurs at a faster rate, 1–2 ms.

The neuropeptide mediates its effects through cell-surface somatostatin receptors (sst). Synthesised and packed in secretion vesicles, these are translocated to the cell membrane. Receptors have seven α-helical transmembrane domains and are coupled to $G_{i/o}$ and $G_{q/11}$ proteins. Six subtypes, $sst_{1/2a,b/3–5}$, have been cloned from humans,

(Patel 1999) but only $sst_{1/2a,b/5}$ are identified in the CNS including: (i) SS-positive PeVN cells (Lanneau *et al* 2000, Breder *et al* 1992, Kumar 2012), (ii) the ArcN-ME circumventricular organ (Tannenbaum *et al* 1998, Breder *et al* 1992, Stroh *et al* 2009), and (iii) somatrotrophs (Yang and Chen 2007, Córdoba-Chacón *et al* 2011). Although the receptors are efficacious when they act independently, they dimerise to boost their functionality (Rocheville *et al* 2000, Kumar 2013). Thus, the sst_1 and sst_5 operating as monomers, demonstrate a lack of internalisation even after prolonged treatment with agonists. However, when dimerizated, their ligand-binding affinity and the ability of heterodimer to phosphorylate, internalise and desensitise is considerably enhanced (Vilardaga *et al* 2010). By contrast, the sst_2 in basal conditions exists as a homodimer which, after exposure to SS, immediately dissociates into monomers.

The biological outcomes of the hormone action depend on the concentration of the neuropeptide, the type of receptor involved, their conditioning and tissue distribution. The binding of SS to $sst_{1/2}$ with the formation of the reactive complex causes: (i) the inhibition of adenylyl cyclase/cAMP/PkA signalling; (ii) the decrease in activity in L-, T- and N-type Ca^{2+} channels and Na^+/K^+ exchanger, (iii) the increase in permeability of A- and M-K^+, delayed and inward rectifier K^+ and Ca^{2+}-activated K^+ ion channels, and (iv) the activation of phosphotyrosine phosphatase, protein kinase G (PkG) and the PLC pathway. As a result, the ultimate effect of the hormone in the body is expected to be inhibitory, e.g. this directly or indirectly inhibits the secretion of GH, GHRH, IGF-1, Ghr and insulin among others. Recent evidence obtained from various tissues and experimental environments indicates that: (i) the stimulation of sst_1 increases whilst sst_2 decreases the sensitivity to Glu cultured hypothalamic neurons, implicating implicitly the excitatory/inhibitory action of the neuropeptide (Lanneau *et al* 1998); (ii) the activation of $sst_{2a,b}$ fosters insulin secretion (Cheng *et al* 2002); (iii) the hormone acting at low concentrations (100 pM–1 μM) excites cortical neurons in culture (Delfs and Dischter 1983); (iv) when applied ionophoretically SS enhances the excitatory effect of ACh on hippocampal and parietal cortex neurons in a dose-dependent manner (Mancillas *et al* 1986); (v) the peptide mediates smooth muscle contraction via the sst_3 (Lahlou *et al* 2004). In addition, using biochemical, confocal and fluorescent recovery imaging approaches, it has been convincingly demonstrated that receptor desensitisation and internalisation play a salient role in the spectrum of SS responses. The desensitisation leads to the dissociation of the SS-$sst_{1/2a,b}$-$G_{\alpha\gamma\beta}$-complex and the termination of the GPCR signalling. During the following stage of internalisation, the receptor is removed from the cell surface. This involves the recruitment of cytosolic β_1–β_2-arrestins, the fusion of remnants of the receptor with clathrin coated pits and their transport to the perinuclear compartment for recycling. During the process, arrestins form $sst_{1/2a,b}$-arrestin-reactive complexes which selectively initiate down-stream intracellular signalling pathways independent from GPCR, and/or serve as a nuclear messenger as evidenced in the case of β_1-arrestin (Schonbrunn 2009, Liu *et al* 2005, Lelouvier *et al* 2008).

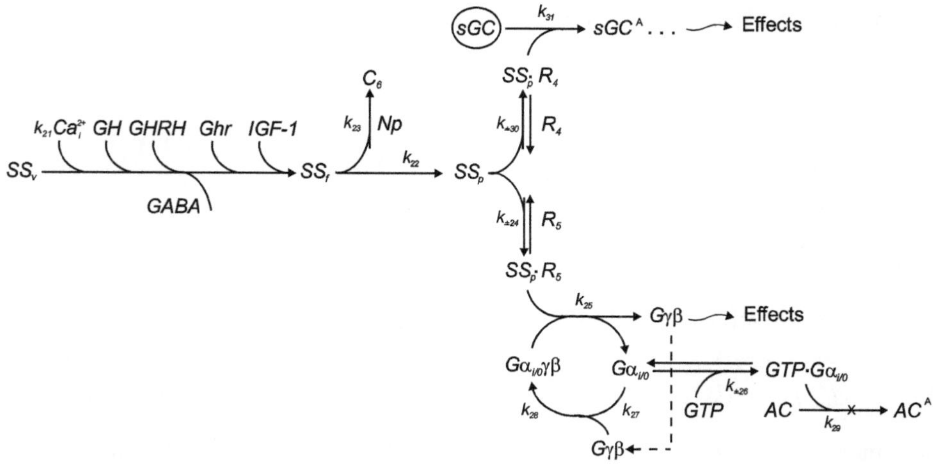

Figure 4.4. Diagram of somatostatin transmission. Here R_4, R_5 are the metabotropic SS type 1 and 2 receptors, respectively.

There are no selective re-uptake mechanisms of SS. Subsequent to secretion, it is rapidly degraded by capillary aminopeptidases and neuro/endopeptidases limiting the half-life of the peptide in plasma to 0.5–3.0 min (Pardridge *et al* 1985).

The general scheme of somatostatin action is given in figure 4.4.

The related system of equations is

$$d\mathbf{X}_S/dt = \mathbf{A}_3\mathbf{X}_S(t) \tag{4.7}$$

where $\mathbf{A}_3$ is the matrix of rate constants and the vector of $\mathbf{X}_S(t) = (x_{27-41})^{\mathrm{T}}$ is

$$\begin{array}{lll}
x_{27} = [SS_{\mathrm{v}}], & x_{28} = [SS_{\mathrm{f}}], & x_{29} = [SS_{\mathrm{p}}] \\
x_{30} = [Np], & x_{31} = [sst_1], & x_{32} = [SS_{\mathrm{p}} \cdot sst_1] \\
x_{33} = [G\alpha_{i/o}\gamma\beta], & x_{34} = [G\alpha_{i/o}]x_{35} = [\gamma\beta], & x_{36} = [GTP \cdot G\alpha_{i/o}] \\
x_{37} = [G], & x_{38} = [G\gamma\beta]x_{39} = [AC], & x_{40} = [sst_1] \\
x_{32} = [SS_{\mathrm{p}} \cdot sst_2]. & &
\end{array}$$

After state-stationary approximations, the system (4.7) can be reduced to

$$d\mathbf{X}_S/dt = \mathbf{A}_3\mathbf{X}_S(t) + \mathbf{C}_S. \tag{4.8}$$

Here the vector of unknowns is $\mathbf{X}_S(t) = (x_{27-30}, x_{33-39})^{\mathrm{T}}$, and the vector $\mathbf{C}_S(0, 0, C_{s(3)}, 0, 0, 0, C_{s(7)}, 0, C_{s(9)}, 0, 0)^{\mathrm{T}}$ has components

$$\begin{aligned}
C_{s(3)} &= k_{-24}x_{32} + k_{-30}x_{41} \\
C_{s(7)} &= k_{35}x_{47}x_{48} + k_{45}x_{57}) \\
C_{s(9)} &= (k_{26}x_{34} + k_{37}x_{49})x_{22}.
\end{aligned}$$

The non-zero elements of the reduced matrix $\mathbf{A}_3(11 \times 11)$ are

$$d_{11} = -d_{21} = -k_{21} g_{\text{syn}}^{\text{Ca}} V^{\text{f}}(t)\left([\text{Ca}_0^{2+}] + x_{53} + x_{56}\right) + k_{-21} x_{86}$$
$$d_{22} = k_{22} + k_{23} x_{30}, \qquad d_{32} = k_{22}$$
$$d_{33} = -(k_{24} x_{31} + k_{30} x_{40}), \qquad d_{44} = -k_{23} x_{28}$$
$$d_{55} = -k_{25} x_{32}, \qquad d_{56} = k_{28} x_{38}$$
$$d_{65} = d_{75} = k_{25} x_{32}, \qquad d_{66} = -(k_{26} x_{22} + k_{28} x_{38})$$
$$d_{68} = k_{-26}, \qquad d_{77} = -k_{27} x_{37}$$
$$d_{86} = k_{26} x_{22}, \qquad d_{88} = -(k_{26} + k_{29} x_{39})$$
$$d_{10,9} = k_{37} x_{37}, \qquad d_{99} = -(k_{27} x_{35} + k_{-26} x_{36} - k_{46} x_{58})$$
$$d_{10,10} = -(k_{28} x_{34} + k_{38} x_{49} + k_{47} x_{58}), \qquad d_{11} = -(k_{29} x_{36} + k_{39} x_{124}).$$

The concentrations of $x_{31,32}$, $x_{40,41}$ are calculated using algebraic equations

$$\begin{aligned}
x_{31}(t) &= \frac{(k_{-24} + k_{25} x_{33}) x_{31}^0}{k_{-24} + k_{24} x_{29} + k_{25} x_{33}} \\
x_{32}(t) &= \frac{k_{24} x_{31}^0 x_{29}}{k_{-24} + k_{24} x_{29} + k_{25} x_{33}} \\
x_{40}(t) &= \frac{(k_{-30} + k_{31} x_{20}) x_{40}^0}{k_{-30} + k_{30} x_{29} + k_{31} x_{20}} \\
x_{41}(t) &= \frac{k_{30} x_{40}^0 x_{29}}{k_{-30} + k_{30} x_{29} + k_{31} x_{20}}.
\end{aligned} \tag{4.9}$$

Given the initial concentrations of the reacting components, $\mathbf{X}_S(0) = \mathbf{X}_{S0}$, equations (4.8)–(4.9) describe the dynamics of neurotransmission by somatostatin.

4.5 Growth hormone releasing hormone

Growth hormone releasing hormone, a peptide of 44 amino acids with a molecular mass of 3358.9 Dalton, plays a key role in body growth and homeostasis over an entire lifespan. GHRH is produced in the CNS and peripheral tissues, e.g. the uterus, kidney, prostate, liver among others (Barabitus and Schally 2010). In the brain, the neuropeptide is synthesised from the pre-pro-GHRH precursor in neurons of the ArcN in the hypothalamus. Following the preprohormone cleavage by signal peptidase, the prohormone is sequestered in nascent vesicles together with processing enzymes such as cathepsin L and subtilisin-like pro-protein convertase. Transported slowly down the axon, inept GHRH undergoes post-translational maturation and packaging in dense granules. Finally, during the reaction of amidation of the C–terminus, the biologically active form of the hormone is produced and stored in LDCVs (100–200 nm in diameter) (Nillni *et al* 1999, Hook *et al* 2008, Russo 2017). By using the wide-field epifluorescence stereo- and confocal laser scanning immunomicroscopy technique, these have been visualised throughout ArcN-ME complex neurons especially in the somas, dendrites and axonal varicosities (Baccam *et al* 2007, Balthazar *et al* 2003). The GHRH-LDCVs often co-localise and co-release with SVs containing GABA using different mechanisms (Hrabovszky *et al* 2005, De-Miguel and Nicholls 2015, Russo 2017).

The observed pattern of GHRH secretion from the ArcN-ME complex is similar and reciprocally related to the dynamics of SS release, i.e. every 15–20 min over an hour with a 3.3 h periodicity (Stroh *et al* 2009). The main stimulus that triggers extrasynaptic exocytosis of GHRH-neurons is prolonged high-frequency electrical depolarisations (see section 4.4 for details). The hormone release can also be induced by Ghr, IGF-1, Glu, and ACh when interacting with their receptors, specifically, Ghr, IGF-1, $GluR_{1/2/3}$, $mGluR_{1a}$ and $NMDA_1$, and $\mu_{1/3}$ (Zigman *et al* 2006, Laviano *et al* 2012, Yin *et al* 2014, Bennett *et al* 1997, Mano-Otagiri *et al* 2006, Osterstock *et al* 2016, Werther *et al* 1989, Petralia and Wenthold 1992, van den Pol *et al* 1994, Jeong *et al* 2016, Kiss *et al* 1997, Kiss *et al* 2006, Baccam *et al* 2007, Gautam *et al* 2009). Ghrelin can also exert its excitatory effect by targeting directly low voltage-activated Ca^{2+} channels. The downregulation of hormone secretion is orchestrated by adrenaline, NPY, GABA and SS through the stimulation of α_2-, NPY_{1-5}, $GABA_A$, $sst_{1/2}$ and, possibly, GH receptors (Breder *et al* 1992, Sun and Miller 1999, Ibata and Watanabe 1977, Baccam *et al* 2007, Zhang and van den Pol 2015, van den Pol *et al* 1994, Stroh *et al* 2009, Zheng *et al* 1997). It is noteworthy that: (i) there exists only indirect evidence of GH receptor presence in GHRH-neurons in the ArcN-ME complex (Lobie *et al* 1993, Hallberg and Nyberg 2012), (ii) SS when co-released on its own or with Glu in PeVN neurons augments/reduces the postsynaptic sensitivity to Glu via sst_1/sst_2 receptors, respectively (Hrabovszky *et al* 2005, Lanneau *et al* 2000), and (iii) NPY hyperpolarizes the cell membrane by increasing K^+ and decreasing Ca^{2+} ion currents (Sun and Miller 1999).

A rough assessment of the dynamics of GHRH release can be made based on available data, namely: the dimension of a LDCV (ves), ~100–200 nm; the concentration of peptide in a vesicle, 3–10 mM; the rate of secretion, 10^3 ves/s, and the number of LDCVs in a neuron, 2×10^5–10^6. The predicted amount of secreted hormone from a vesicle is $\sim(0.8–2) \times 10^{-2}$ $fM \cdot s^{-1}$. Assuming that half of all LDCVs are released from a single neuron over 5 s, the projected concentration of GHRH will yield $(0.4–1) \times 10^{-1}$ pM. If it is further assumed that 10^3 neurons are simultaneously involved, then the estimated total amount of secreted hormone yields 0.04–0.1 nM.

GHRH is released into the hypophyseal portal blood system in a pulsatile manner. Once the neuropeptide is delivered through blood flow to the adenohypophysis, it binds to G_s-protein-coupled cell surface receptors (GHRH) on somatotrophs. The signal transduction involves the activation of Na^+, L- and T-type Ca^{2+} channels, adenylate cyclase/cAMP/PkA and, possibly, PLC pathways leading to upregulation of the somatotrophs' mitogenesis, their gene transcription, GH *de novo* production and secretion (Pombo *et al* 2000, Mayo *et al* 2000, Anderson *et al* 2004, Matsoukas and Spyroulias 2017).

As is the case with any peptide, there is no re-uptake mechanism for GHRH. High performance liquid chromatography, radioimmunoassay and bioassay *in vitro* and *in vivo* studies of the metabolic clearance and elimination rates of hormone by

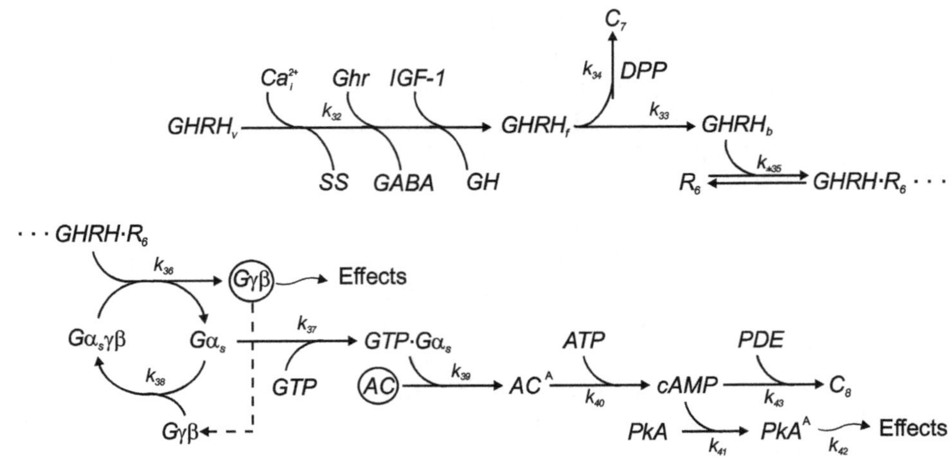

Figure 4.5. State diagram of GHRH transmission. Here R_6 is the GHRH receptor.

plasma dipeptidylaminopeptidase have revealed a half-time degradation $t_{1/2} = 2.7 \pm 0.6$ min and a disappearance rate of $t_{1/2} = 16 \pm 1.2$ min (Frohman *et al* 1986).

The diagram of GHRH conversion is shown in figure 4.5.

The system of equations that corresponds to the sequence of reactions as shown is

$$d\mathbf{X}_{GHRH}/dt = \mathbf{A}_4\mathbf{X}_{GHRH}(t) \tag{4.10}$$

where the components of the vector $\mathbf{X}_{GHRH}(t) = (x_{42-53}, x_{124}, x_{125})^{\mathrm{T}}$ are

$$\begin{array}{lll} x_{42} = [GRHR_{\mathrm{v}}], & x_{43} = [GHRH_{\mathrm{f}}], & x_{44} = [GHRH_{\mathrm{p}}] \\ x_{45} = [DPP], & x_{46} = [GHRH \cdot R], & x_{47} = [GHRH_{\mathrm{p}} \cdot GHRH \cdot R] \\ x_{48} = [G\alpha_s\gamma\beta], & x_{49} = [G\alpha_s], & x_{50} = [ATP] \\ x_{51} = [cAMP], & x_{52} = [PkA], & x_{53} = [PkA^{\mathrm{A}}] \\ x_{124} = [GTP \cdot G\alpha_s], & x_{125} = [AC^{\mathrm{A}}]. & \end{array}$$

The reduced form of the system (4.10) yields

$$d\mathbf{X}_{GHRH}/dt = \mathbf{A}_4\mathbf{X}_{GHRH}(t) + \mathbf{C}_{GHRH}. \tag{4.11}$$

Here $\mathbf{X}_{GHRH}(t) = (x_{42-45}, x_{48-51}, x_{124}, x_{125})^{\mathrm{T}}$, $\mathbf{C}_{GHRH}(0, 0, k_{-35}[x_{47}],0,\&,0)^{\mathrm{T}}$, and the matrix $\mathbf{E}(10 \times 10)$ has the following non-zero elements

$$e_{11} = -e_{21} = -k_{32}g_{\mathrm{syn}}^{\mathrm{Ca}}V^{\mathrm{f}}(t)([\mathrm{Ca}_0^{2+}] + x_{56}) + k_{-32}(x_{25} + x_{38})$$

$$\begin{array}{lll} e_{22} = -(k_{33} + k_{34}x_{45}), & e_{32} = k_{33}, & e_{33} = -k_{35}x_{46} \\ e_{44} = -k_{34}x_{44}, & e_{55} = -k_{36}x_{47}, & e_{56} = k_{38}x_{38} \\ e_{65} = k_{36}x_{47}, & e_{66} = -(k_{37}x_{22} + k_{38}x_{38}) & \\ e_{77} = -e_{87} = -k_{40}x_{125}, & e_{88} = -(k_{41}x_{52} + k_{43}x_{26}), & e_{96} = k_{37}x_{22} \\ e_{99} = -e_{10,9} = -k_{39}x_{39}, & e_{10,10} = -k_{40}x_{50}. & \end{array}$$

The dynamics of $x_{46,47}$, $x_{52,53}$ is given by

$$
\begin{aligned}
x_{46}(t) &= \frac{(k_{-35} + k_{36}x_{48})x_{46}^{0}}{k_{-35} + k_{35}x_{44} + k_{36}x_{48}} \\
x_{47}(t) &= \frac{k_{35}x_{46}^{0}x_{44}}{k_{-35} + k_{35}x_{44} + k_{36}x_{48}} \\
x_{52}(t) &= \frac{k_{42}x_{52}^{0}}{k_{42} + k_{41}x_{51}} \\
x_{53}(t) &= \frac{k_{41}x_{52}^{0}x_{51}}{k_{42} + k_{41}x_{51}}.
\end{aligned}
\tag{4.12}
$$

Given the initial concentrations of the reacting components, $\mathbf{X}_{GHRH}(0) = \mathbf{X}_{GHRH0}$, equations (4.11) and (4.12) describe the dynamics of GHRH signalling.

4.6 Ghrelin

Ghrelin, a peptide of 28-amino acids with a molecular mass of 3370.9 Dalton, is synthesised mainly by mucosal P/D cells of oxyntic glands in the fundus of the stomach although other sites including the hypothalamus, organs of the gastrosintestinal tract, pancreas, kidneys, etc are implicated (Ariyasu *et al* 2001, Sato *et al* 2012). The hormone is produced from preproghrelin, a 117-amino acid, through multiple stages of cleavage and post-translational acylation at the serine 3 residue by ghrelin *O*-acyltransferase (Yang *et al* 2008). The acylated fraction, Ghr, exhibits a broad range of biological activity, whilst the non-acylated form, desacyl-Ghr, lacks neuroendocrine efficacy (Brogio *et al* 2004, Guillemin 2005). The concentration of the two fractions in human plasma is 10–20 fM ml^{-1} and 90–130 fM ml^{-1}, respectively, an amount which fluctuates depending on the fasting (increased) or feeding (reduced) state.

There are two isoforms of the Ghr receptor ($GhrR_{1a/1b}$) which belong to the family of $G_{\alpha q/11}$-coupled receptors (Yin *et al* 2014, Laviano *et al* 2012). Their distribution in the CNS has been assessed by *in situ* hybridisation histochemistry. The most abundant and strongest GhrR signals have been recorded from the ArcN (Zigman *et al* 2006). GhrR are also expressed in the anterior pituitary gland (somatotrophs), anteroventral PeVN, substantia nigra, dorsal and medial raphe nuclei, ventral tegmental area and hippocampus implying the multifaceted functionality of Ghr, e.g. adenohypophysis hormones secretion, appetite and mood control, glucose metabolism, etc (Reichenbach *et al* 2012, Guan *et al* 1997, Korbonits *et al* 2001a, 2001b). Ghr elicits its effects through binding to $GhrR_{1a}$ type either by the activation of the intracellular PLC pathway or by targeting directly L-, N-type voltage-dependent transmembrane Ca^{2+} channels (Yin *et al* 2014). By contrast, $GhrR_{1b}$ type does not bind to the peptide but rather attenuates the trafficking of $GhrR_{1a}$ to the cell surface decreasing its constitutive activity.

In the brain Ghr modulates the secretion of GHRH from the ArcN-ME organ and GH from somatotrophs. Quantitatively, the effect of Ghr on GH release is 2–3 times greater when compared to GHRH. Acting synergistically, the cumulative hormonal response is higher than the sum of the effects of GHRH and Ghr acting separately (Gnanapavan *et al* 2002, Hataya *et al* 2001, Arvat *et al* 2001).

The degradation of peptide results from the reaction of desoctanoylation by butyrylcholinesterase or dipeptidylpeptidase without proteolysis. The biological half-life of acylated hormone is 9–13 min whilst the elimination of the total amount of Ghr is $t_{1/2}$ ~30 min (Akamizu *et al* 2004).

The general scheme of ghrelin action in the body is given in figure 4.6. The overall dynamics of chemical reactions is given by the system

$$d\mathbf{X}_{Ghr}/dt = \mathbf{A}_5\mathbf{X}_{Ghr}(t) \tag{4.13}$$

where $\mathbf{A}_5$ is the matrix of rate coefficients and the vector $\mathbf{X}_{Ghr}(t) = (x_{54-69})^{\mathrm{T}}$ is defined by

$$\begin{array}{llll} x_{54} = [Ghr], & x_{55} = [Ghr \cdot R], & x_{56} = [Ghr \cdot Ghr \cdot R], & x_{57} = [G\alpha_q\gamma\beta] \\ x_{58} = [G\alpha_q], & x_{59} = [GTP \cdot G\alpha_q], & x_{60} = [PLC], & x_{61} = [PLC^{\mathrm{A}}] \\ x_{62} = [PIP_2], & x_{63} = [IP_3], & x_{64} = [DAG], & x_{65} = [PkC] \\ x_{66} = [PkC^{\mathrm{A}}], & x_{67} = [Ca_{iER}^{2+}], & x_{68} = [GTP \cdot Ca_{iER}^{2+} \cdot PkC], & x_{69} = [DGAT] \end{array}$$

Equation (4.13) in their reduced matrix form are

$$d\mathbf{X}_{Ghr}/dt = \mathbf{A}_5\mathbf{X}_{Ghr}(t) + \mathbf{C}_{Ghr} \tag{4.14}$$

here $\mathbf{X}_{Ghr} = (x_{54}, x_{57-59}, x_{64}, x_{67}, x_{69})^{\mathrm{T}}$, $\mathbf{C}_{Ghr}(C_{Ghr(1)}, 0, C_{Ghr(3)}, 0, C_{Ghr(5)}, C_{Ghr(6)}, 0)^{\mathrm{T}}$ with $C_{Ghr(1)} = k_{-44}x_{56}$, $C_{Ghr(3)} = k_{49}x_{61}x_{62}$, $C_{Ghr(5)} = C_{Ghr(3)} + k_{-51}x_{66}$, $C_{Ghr(6)} = (k_{-52} + k_{53})x_{68}$. The reduced matrix $\mathbf{A}_5(7 \times 7)$ has the following non-zero elements

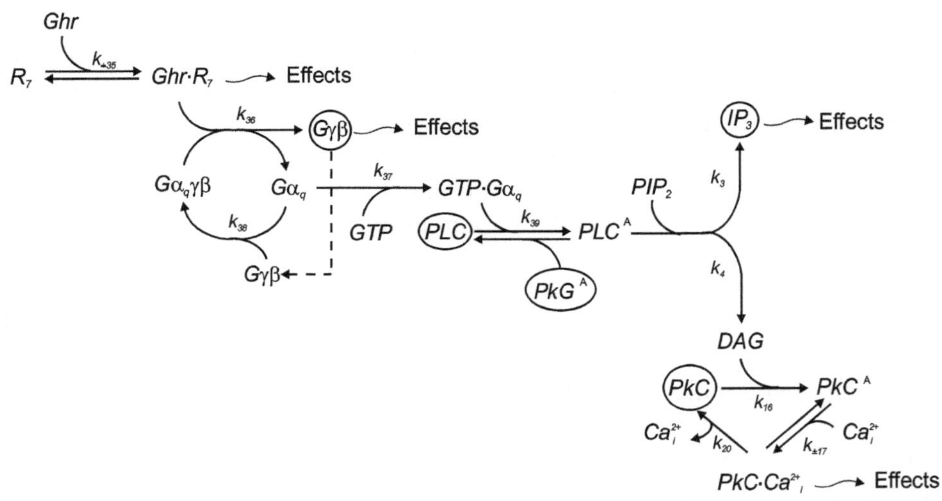

Figure 4.6. State diagram for Ghr signalling. Here R_7 is the Ghr receptor.

$$f_{11} = -k_{44}x_{54}, \qquad f_{23} = k_{47}x_{38}, \qquad f_{22} = -f_{32} = -k_{45}x_{56}$$
$$f_{33} = -(k_{46}x_{22} + k_{47}x_{38}), \qquad f_{43} = k_{46}x_{22}, \qquad f_{44} = -k_{48}x_{60}$$
$$f_{55} = -(k_{51}x_{65} + k_{54}x_{69}), \qquad f_{66} = -k_{52}x_{66}, \qquad f_{77} = -k_{54}x_{64}.$$

The substrates $x_{55,56}$, x_{60-62}, $x_{65,66}$, x_{68} are in a quasi-steady state, and are calculated from

$$\begin{aligned}
x_{55}(t) &= \frac{(k_{-44} + k_{45}x_{57})x_{55}^{0}}{k_{-44} + k_{44}x_{54} + k_{45}x_{57}} \\
x_{56}(t) &= \frac{k_{44}x_{55}^{0}x_{54}}{k_{-44} + k_{44}x_{54} + k_{45}x_{57}} \\
x_{60}(t) &= \frac{(k_{50}x_{25} + k_{69}x_{62})x_{60}^{0}}{k_{48}x_{59} + k_{49}x_{62} + k_{50}x_{25}} \\
x_{61}(t) &= \frac{k_{48}x_{60}^{0}x_{59}}{k_{48}x_{59} + k_{49}x_{62} + k_{50}x_{25}} \\
x_{62}(t) &= x_{62}^{0} - x_{63} \\
[x_{63}](t) &= \frac{k_{49}x_{61}x_{62}x_{126}}{k_{98} + x_{126}} \\
x_{65}(t) &= \frac{k_{-51} + k_{51}x_{67}}{k_{-51} + k_{51}x_{64} + k_{52}x_{67}}\left[x_{65}^{0} - \frac{k_{-51}x_{68}}{k_{-51} + k_{52}x_{67}} - \frac{k_{52}x_{66}x_{67}}{k_{-52} + k_{53}}\right] \\
x_{66}(t) &= \frac{k_{51}x_{64}x_{65} + k_{-52}x_{68}}{k_{-51} + k_{52}x_{67}} \\
x_{68}(t) &= \frac{k_{52}x_{66}x_{67}}{k_{-52} + k_{53}}.
\end{aligned} \tag{4.15}$$

Given the initial concentrations of the reacting components $\mathbf{X}_{Ghr}(0) = \mathbf{X}_{Ghr0}$, equations (4.14) and (4.15) describe the dynamics of Ghr signalling.

4.7 Growth hormone

Growth hormone, a polypeptide of 191-amino acids with a molecular weight of 22.124 kDa, is synthesised by somatotrophs at a rate of 327 ± 164 μg/24 h (men) and 392 ± 157 μg/24 h (women) (Winer *et al* 1990). The prehormone, a 217-amino-acid, is cleaved to 1–26-amino-acid peptide (mature GH) and stored in LDCVs dispersed throughout the cell. There are ~(4.9 ± 0.21)/μm^2 fully filled and ~(0.5–1.2)/μm^2 partly filled or empty vesicles in porcine somatotrophs (Lee *et al* 2004, Kurosumi *et al* 1986, Sasaki 1987). Combining these with the cell morphometric data, it can be estimated that ~520 lDCVs containing ~0.19 fM ml^{-1} of the peptide are present in the porcine anterior pituitary gland. At the time of writing no data was available for humans.

The secretion of GH is upregulated by GHRH, Ghr, ACh, AD, ATP and dopamine through their respective GHRH-R, GhrR, μ_3, α_2, $P2X_2$ and dopamine receptors. On stimulation, these initiate a cascade of intracellular processes leading to depolarisation of the membrane, a rise in cAMP and cytosolic Ca^{2+}, and, eventually, extrasynaptic exocytosis of the hormone (Lussier *et al* 1991a, 1991b, 1991c, Bjelobaba *et al* 2015, Rawlings and Hezareh 1996, Zemkova and Stojilkovic 2018, Gautam *et al* 2009). The actual mechanism of the release is similar to GHRH and SS. Other peptides are produced in the hypothalamus and periphery, e.g. thyrotropic releasing hormone, pituitary adenylate cyclase-activating peptide, leptin, corticotropin-releasing hormone, galain, and NPY, to regulate the secretion of GH. However, these are beyond the scope of the current book (Anderson *et al* 2004).

Downregulation of GH secretion is under the direct control of SS, endothelin, GABA, and AD, and may be affected indirectly by GH itself through NPY neurons in the ArcN. These foster their effects through IGF-1R, $sst_{1/2}$, ET_A, $GABA_A$ and β_2 receptors, and associated signalling pathways (Kineman *et al* 2018, Sugihara *et al* 1999, Minami *et al* 1998, Bonert and Melmed 2017, Tomić *et al* 1999, Lussier *et al* 1991c). Evidence for the presence of GH-Rs on somatotrophs is inconclusive, arguing for possible autocrine modulation of the polypeptide release (Asa *et al* 2000, Hallberg and Nyberg 2012). IGF-1 suppresses GH synthesis leaving the hormone secretion unaffected. This occurs through decreasing GH gene expression by the phosphorylation of CREP binding protein CBP and disrupting its binding to the transcription factor Pit-1 in the promoter region (Romero *et al* 2012, Gahete *et al* 2013).

GH release in humans follows an ultradian rhythm, being both sporadic and unique to each individual. Ten to fifteen 'bursts' occur throughout the day. On average $\sim 10.0 \pm 1.6\ \mu g\ h^{-1}$ of the hormone is secreted into the blood circulation during the daylight hours with a maximum $\sim 28.5 \pm 2.5\ \mu g\ h^{-1}$ surge in the second half of the night (Alford *et al* 1973). The pulsatility is reciprocally linked to the production and secretion of GHRH and SS. The former influences the amount, whilst the latter regulates the timing of GH exocytosis. Once released, nearly half of the peptide is quickly bound to the high affinity specific GH-binding serum protein (GHBP) (Vijayakumar *et al* 2011, Schilbach and Bidlingmaier 2015). This mechanism prevents acute oscillation in levels, stabilises and prolongs a plasma half-life ($t_{1/2}$ = 15–51 min), and regulates the bioavailability of the hormone (Schaffer *et al* 1996).

GH elicits its biological effects by binding to growth hormone receptors (GH-Rs). These belong to the class 1 cytokine receptor family (Dehkhoda *et al* 2018, Bonert and Melmed 2017). They exist as natural homodimers and structurally consist of three domains, namely the extra-, transmembrane, and intracellular. The extracellular domain is a ligand-binding region whilst the cytosolic domain is closely associated with the tyrosine kinase Janus kinase (JAK2) which serves as the functional 'switch' for the receptor (Waters 2015, Waters *et al* 2014, Brooks *et al* 2014, Herington and Lobie 2012, Vijayakumar *et al* 2011, Bernat *et al* 2003). The binding of GH to the receptor followed by conformational changes, triggers the activation of the JAK-dependent and/or JAK-independent signalling cascades. The former involves the down-stream phosphorylation of JAK2 with a subsequent recruitment of signal transducers and activators of transcription

($STAT_{1-4/5a,b/6}$) proteins, *c-fos*, insulin receptor substrate 1 (IRS-1), and focal adhesion kinase (Barclay *et al* 2010, Fernandez-Pérez *et al* 2016, Dehkhoda *et al* 2018). These function as transduction factors in GH-induced reactions. Thus, once activated: (i) $STAT_{5a,b}$ translocate from the cell surface to the nucleus where they bind to nuclear DNA and elicit GH-specific target gene expression, e.g. IGF-1 transcription (Chia 2014); (ii) IRS-1 stimulates key proteins, phosphatidylinositol 3-kinase (PI3K) and protein kinase B (PkB), these being the main elements of the PI3K/PkB pathway which up-regulates myogenesis and collagen synthesis; (iii) focal adhesion kinase leads to the reorganisation of the actin cytoskeleton promoting cell differentiation (Herington and Lobie 2012). By contrast, the JAK-independent pathway leads to direct activation of STATs and Src family tyrosine kinases (SFK). The phosphorylation of non-receptor SFK, the activation of phospholipase Cγ, Ras-guanine nucleotide exchange factor (Ras), mitogen activated (MAPK) extracellular signal-regulated (ERK-1/2) kinases—the PLCγ/Ras/MAPK/ERK1–2 pathway—instigate cell growth and proliferation (Rowlinson *et al* 2008).

Inactivation of GH-Rs is mediated through the suppression of cytokine signalling ($SOCS_{1/2/3/6}$) and tyrosine phosphatase proteins. Both target JAKs, GH-Rs, and the protein inhibitor of activated STAT. SOCS proteins play a major role in the negative feedback loop of GH signalling. They inhibit the ability of JAK2 and STATs to bind to GH-Rs, facilitating their ubiquitination for degradation (Linossi *et al* 2013, Pilecka *et al* 2007). The function of the protein inhibitor of the activated STAT still requires further elucidation. It has been suggested that it operates by directly preventing STATs-DNA binding and, possibly, by recruiting other transcriptional co-repressors. GH-Rs *per se* also undergo proteolytic cleavage, the process catalysed by the tumor necrosis factor converting enzyme, which downregulates GH-R sensitivity and ultimately signalling (Frank and Fuchs 2008) (figure 4.7).

The system of equations for GH conversion is given by

$$d\mathbf{X}_{GH}/dt = \mathbf{A}_6\mathbf{X}_{GH}(t) \tag{4.16}$$

where $\mathbf{A}_6$ is the matrix of rate constants and the vector $\mathbf{X}_{GH}(t) = (x_{70-90})^{\mathrm{T}}$ is defined by

$$x_{70} = [GH_{\mathrm{v}}], \quad x_{71} = [GH_{\mathrm{f}}], \quad x_{72} = [GH_{\mathrm{b}}], \quad x_{73} = [GHBP_3]$$
$$x_{74} = [GH_{\mathrm{f}} \cdot GHBP_3], \quad x_{75} = [GH \cdot R \cdot JAK2], \quad x_{76} = [GH \cdot R \cdot JAK2^{\mathrm{A}}]$$
$$x_{77} = [STAT5_{\mathrm{a,b}}], \quad x_{78} = [STAT5^{\mathrm{A}}{}_{\mathrm{a,b}}], \quad x_{79} = [SOCS2]$$
$$x_{80} = [SOCS2^{\mathrm{A}}], \quad x_{81} = [IRS - 1], \quad x_{82} = [IRS - 1^{\mathrm{A}}]$$
$$x_{83} = [DNA], \quad x_{84} = [DNA^{Tr}], \quad x_{85} = [GH \cdot R]$$
$$x_{86} = [GH_{\mathrm{b}} \cdot GH \cdot R], \quad x_{87} = [SFK], \quad x_{88} = [GH_{\mathrm{b}} \cdot GH \cdot R \cdot SFK]$$
$$x_{89} = [PLC\gamma], \quad x_{90} = [PLC^{\mathrm{A}}\gamma].$$

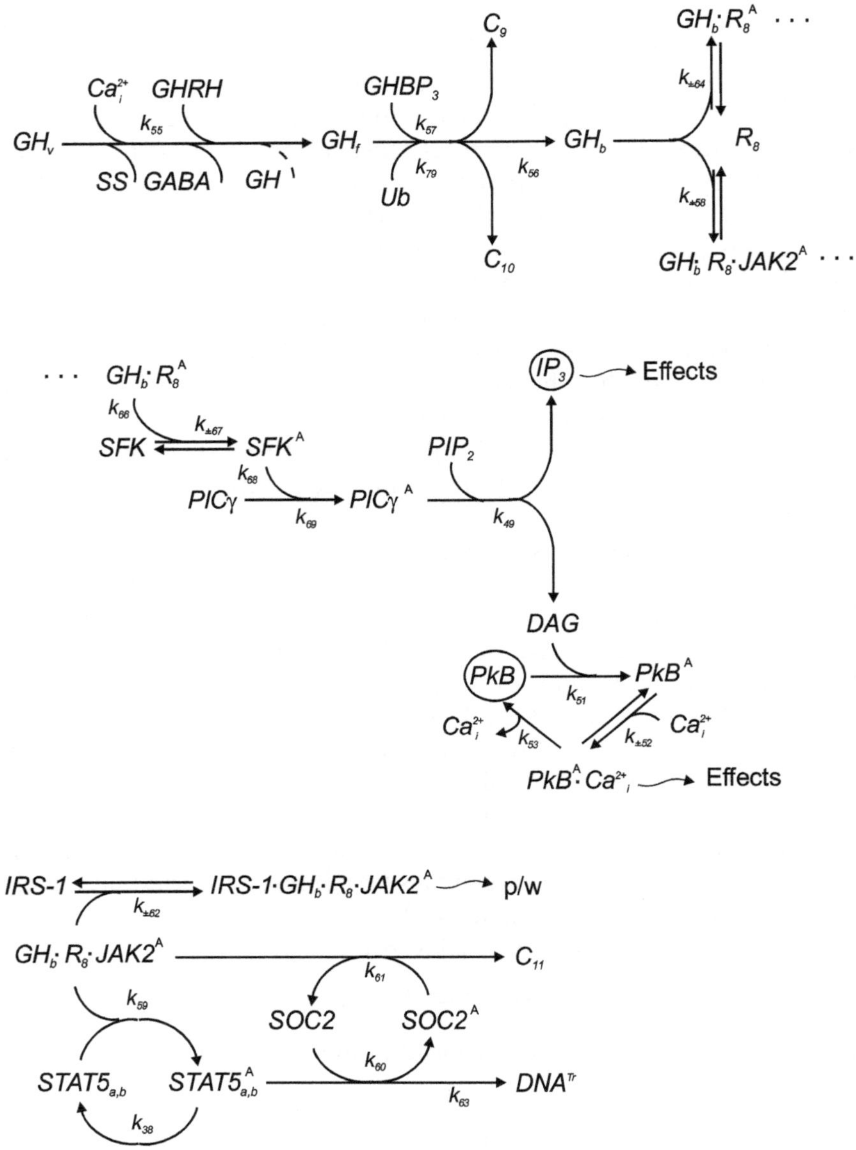

Figure 4.7. Diagram of GH signalling. Here R_8 is the GH receptor.

The system (4.16) in its reduced form is

$$d\mathbf{X}_{GH}/dt = \mathbf{A}_6\mathbf{X}_{GH}(t) + \mathbf{C}_{GH} \tag{4.17}$$

where $\mathbf{X}_{GH}(t) = (x_{70-72}, x_{74-76}, x_{83}, x_{84}, x_{89}, x_{90})^{\mathrm{T}}$, $\mathbf{C}_{GH} = (0, \&, 0, C_{GH(6)}, 0, \&, 0)^{\mathrm{T}}$ with $C_{GH(6)} = k_{-62}x_{82} + k_{-76}x_{98}$, and the matrix $\mathbf{A}_6(10 \times 10)$ is given by

$$g_{11} = -g_{21} = -k_{55} g_{syn}^{Ca} V^f(t)\left([Ca_0^{2+}] + x_{56}\right) + k_{-55}(x_{25} + x_{38})$$
$$g_{22} = -(k_{56} + k_{57}(x_{73}^0 - x_{74}), \quad g_{32} = k_{56}$$
$$g_{33} = -\left(k_{58}x_{75} + \frac{k_{65}k_{66}x_{85}^0}{k_{-65} + k_{66} + k_{65}x_{72}}\right), \quad g_{36} = g_{56} = k_{-58}$$
$$g_{42} = k_{57}x_{73}^0, \; g_{44} = -k_{57}x_{73}^0 x_{71}, \quad g_{55} = -g_{65} = -k_{58}x_{72}$$
$$g_{66} = -(k_{-58} + k_{59}x_{77} + k_{61}x_{80} + k_{62}x_{81} + k_{76}x_{97}),$$
$$g_{77} = -k_{63}x_{78}, \quad g_{87} = k_{63}x_{78}, \quad g_{88} = -k_{70}.$$

As a result of the cross-talk between GH and Ghr at the point of activation of the PLC/PLCγ pathways, the mechanism of conversion of the substrates X_{89}, X_{90} is similar to the chemical kinetics of $x_{60/61}$ studied above.

The variables, x_{73}, x_{77-82}, x_{85-88}, quickly achieve the quasi-equilibrium state and are found from

$$\begin{aligned}
x_{73}(t) &= x_{73}^0 - x_{74} \\
x_{77}(t) &= \frac{(k_{60}x_{79} + k_{63}x_{83})x_{77}^0}{k_{59}x_{76} + k_{60}x_{79} + k_{63}x_{83}} \\
x_{78}(t) &= \frac{k_{59}x_{77}^0 x_{76}}{k_{59}x_{76} + k_{60}x_{79} + k_{63}x_{83}} \\
x_{79}(t) &= \frac{(k_{61}x_{76} + k_{64}x_{77})x_{79}^0}{k_{60}x_{78} + k_{61}x_{76} + k_{64}x_{77}} \\
x_{80}(t) &= \frac{k_{60}x_{79}^0 x_{78}}{k_{60}x_{78} + k_{61}x_{76} + k_{64}x_{77}} \\
x_{81}(t) &= \frac{(k_{-62} + k_{-77} + k_{80})x_{81}^0 + k_{80}x_{100}}{k_{-62} + k_{-77} + k_{62}x_{76} + k_{77}x_{98} + k_{80}x_{100}} \\
x_{82}(t) &= \frac{(k_{62}x_{76} + k_{77}x_{98})x_{81}^0}{k_{-62} + k_{-77} + k_{62}x_{76} + k_{77}x_{98} + k_{80}x_{100}} \\
x_{85}(t) &= \frac{(k_{-65} + k_{66})x_{85}^0}{k_{-65} + k_{66} + k_{65}x_{72}} \\
x_{86}(t) &= \frac{k_{65}x_{85}^0 x_{72}}{k_{-65} + k_{66} + k_{65}x_{72}} \\
x_{87}(t) &= \frac{(k_{-67} + k_{68} + k_{69}x_{60})x_{87}^0}{k_{-67} + k_{68} + k_{67}x_{86} + k_{69}x_{60}} \\
x_{88}(t) &= \frac{k_{67}x_{87}^0 x_{86}}{k_{-67} + k_{68} + k_{67}x_{86} + k_{69}x_{60}}.
\end{aligned} \tag{4.18}$$

Provided the initial concentrations of the substrates, $\mathbf{X}_{GH}(0) = \mathbf{X}_{GH0}$, are known, the system (4.17) together with equation (4.18) describe the dynamics of GH signalling.

Although GH acting alone regulates multiple biological functions in the body, its effects related to growth-promoting activities are primarily mediated by IGF-1.

4.8 Insulin-like growth factor-1

IGF-1, a peptide of 70 amino acids with a molecular weight of 7.649 kDa, is synthesised mainly (75%) by hepatocytes in the liver, with the remaining 25% being produced in response to GH stimulation in practically all the cells and tissues of the human body (Devesa *et al* 2016). The peptide is stored in LDCVs and released into the blood circulation on demand in Ca^{2+} dependent manner together with IGF-binding proteins 1–6 ($IGFBP_{1-6}$) and the acid labile subunit (ALS), a liver-derived glycoprotein (Cao *et al* 2011, Schmidt and Moore 1994). $IGFBP_3$ is the largest, with a molecular weight of 28.717 kDa, and also the most abundant in serum, having a high affinity for IGF-1 (Kim *et al* 1997). It binds nearly 99% of the total IGF—to form the ternary IGF-1-$IGFBP_3$-ALS-complex leaving <1% unbound, 0.1–0.4 nM. The complex modulates the peptide's bioavailability by protecting it from proteolysis, stabilising and sequestering the peptide in the vascular space to extend its half-life time, $t_{1/2} \simeq 10$ min, and acting as a competitive antagonist to the IGF-1 receptor (IGF-1R) (Kineman *et al* 2018). The bioactive peptide is released gradually from the complex through proteolytic cleavage reactions (Laursen *et al* 2007).

The synthesis and release of IGF-1 from hepatocytes is downregulated by GHRH by reversing the activation of growth hormone JAK2-dependent STAT signalling pathway (Cui and Schally 2018).

The human IGF-1R is a disulphide-linked homodimer $(\alpha\beta)_2$ of single-pass integral membrane protein. The two α-extracellular regions possess ligand-binding sites. The two β-subunits are composed of three domains: a short extracellular, a transmembrane and an intracellular. The cytosolic domain is associated with tyrosine kinase and constitutes the signal transduction region (Duan *et al* 2010, Bentov and Werner 2013, Xu *et al* 2018). Following the binding of the peptide, the IGF-1R undergoes a conformational change, leading to autophosphorylation (Kavran *et al* 2014). This in turn induces phosphorylation of tyrosine coupled adaptor substrates including IRS-1, Shc, and STATs. Once activated: (i) IRS-1 triggers the PI3K/PkB pathway; (ii) Shc recruits Ras and prompts the Ras/MAPK/ERK-pathway, and (iii) JAK2 activates directly $STAT_{5a,b}$ proteins. The above pathways can also be stimulated by GH and IGF-1 to form reactive complexes, e.g. GH-R/JAK2/IGF-1R that induce a cascade of intracellular signalling (Herington and Lobie 2012) (figure 4.8).

The mathematical model of IGF-1 action is given by

$$d\mathbf{X}_{IGF}/dt = \mathbf{A}_7\mathbf{X}_{IGF}(t) \tag{4.19}$$

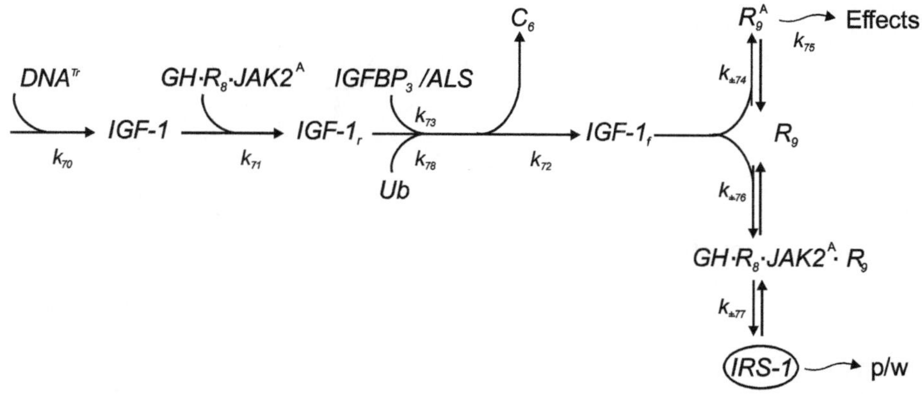

Figure 4.8. Scheme of signal transduction by IGF-1. Here R_9 is the IGF-1 receptor.

where $\mathbf{A}_7$ is the matrix of rate constants and the vector $\mathbf{X}_{IGF}(t) = (x_{91-108})^{\mathrm{T}}$ is

$$
\begin{aligned}
&x_{91} = [IGF1_{\mathrm{v}}], \quad x_{92} = [IGF1_{\mathrm{f}}], \quad x_{93} = [IGF1_{\mathrm{b}}]\\
&x_{94} = [IGFBP_3/ALS], \quad x_{95} = [IGF_{\mathrm{f}} \cdot GHBP_3/ALS], \quad x_{96} = [IGF1 \cdot R]\\
&x_{97} = [IGF1 \cdot R^{\mathrm{A}}], \quad x_{98} = [GH_{\mathrm{b}} \cdot GHR \cdot JAK2^{\mathrm{A}} \cdot IGF1 \cdot R^{\mathrm{A}}]\\
&x_{99} = [Ub], \quad x_{100} = [PI3K], \quad x_{101} = [PI3K^{\mathrm{A}}]\\
&x_{102} = [PIP_3], \quad x_{103} = [PkB], \quad x_{104} = [PkB^{\mathrm{A}}]\\
&x_{105} = [S6K1], \quad x_{106} = [S6K1^{\mathrm{A}}], \quad x_{107} = [S6]\\
&x_{108} = [S6^{\mathrm{A}}].
\end{aligned}
$$

The system (4.19) in its reduced form yields

$$
d\mathbf{X}_{IGF}/dt = \mathbf{A}_7\mathbf{X}_{IFG}(t) + \mathbf{C}_{IGF}. \tag{4.20}
$$

Here the vector of unknowns $\mathbf{X}_{IGF}(t) = (x_{91-93}, x_{95}, x_{98}, x_{99}, x_{102})^{\mathrm{T}}$ and the non-zero components of the vector $\mathbf{C}_{IGF} = (C_{IGF(1)}, 0, C_{IGF(3)}, 0, C_{IGF(5)}, 0, C_{IGF(7)})^{\mathrm{T}}$ are $C_{IGF(1)} = k_{70}x_{84}$, $C_{IGF(3)} = k_{-77}x_{82}$, $C_{IGF(5)} = k_{-74}x_{97}$, $C_{IGF(7)} = k_{81}x_{62}x_{101}$. The reduced matrix $\mathbf{A}_7(7 \times 7)$ has the non-zero elements

$$
\begin{aligned}
&h_{11} = -h_{21} = -k_{71}x_{86}, \quad h_{22} = -(k_{72} + k_{73}(x_{94}^{0} - x_{95}))\\
&h_{26} = k_{78}, \quad h_{32} = k_{72} \quad h_{33} = -k_{74}x_{96}\\
&h_{42} = k_{73}x_{94}^{0}, \quad h_{44} = k_{73}x_{92}, \quad h_{54} = k_{76}k_{77}\\
&h_{55} = -k_{-76} - k_{77}x_{81}, \quad h_{66} = -(k_{78}x_{92} + k_{79}x_{79}), \quad h_{77} = -k_{80}x_{103}.
\end{aligned}
$$

The concentrations of the substrates x_{94}, $x_{96,97}$, $x_{100,101}$, $x_{104-108}$ are found from the algebraic equations

$$
\begin{aligned}
x_{94}(t) &= x_{94}^{0} - x_{95} \\
x_{96}(t) &= \frac{(k_{-74} + k_{75} + k_{76}x_{86})x_{96}^{0} - k_{-76}x_{98}}{k_{-74} + k_{75} + k_{74}x_{93} + k_{76}x_{76}} \\
x_{97}(t) &= \frac{k_{74}x_{93}x_{96} + k_{-76}x_{98}}{k_{-74} + k_{75} + k_{76}x_{76}} \\
x_{100}(t) &= \frac{k_{81}x_{100}^{0}x_{62}}{k_{80}x_{82} + k_{81}x_{62}} \\
x_{101}(t) &= \frac{k_{80}x_{100}^{0}x_{82}}{k_{80}x_{82} + k_{81}x_{62}} \\
x_{103}(t) &= \frac{k_{83}x_{103}^{0}x_{105}}{k_{82}x_{102} + k_{83}x_{105}} \\
x_{104}(t) &= \frac{k_{82}x_{103}^{0}x_{102}}{k_{82}x_{102} + k_{83}x_{105}} \\
x_{105}(t) &= \frac{k_{84}x_{105}^{0}x_{107}}{k_{83}x_{104} + k_{84}x_{107}} \\
x_{106}(t) &= \frac{k_{83}x_{105}^{0}x_{104}}{k_{83}x_{104} + k_{84}x_{107}} \\
x_{107}(t) &= \frac{k_{85}x_{107}^{0}x_{109}}{k_{84}x_{106} + k_{85}x_{109}} \\
x_{108}(t) &= \frac{k_{84}x_{107}^{0}x_{106}}{k_{84}x_{106} + k_{85}x_{109}}.
\end{aligned}
\tag{4.21}
$$

Given the initial concentrations of the substrates, $\mathbf{X}_{IGF}(0) = \mathbf{X}_{IGF0}$, equations (4.20) and (4.21) describe the dynamics of IGF-1signalling.

IGF-1 is a powerful mitogen, stimulating cell proliferation and differentiation when acting either alone or in conjunction with GH. These are essential for muscle and connective tissue growth, hypertrophy and regeneration, all of which are necessary adaptive physiological processes for a constantly changing environment.

4.9 Myogenesis

Myogenesis is a tightly gene-controlled sequence of events. Activation of the myocyte enhancer factor 2 (Mef2) protein by S6k triggers a proliferation of myoblasts (m_b), precursors for muscle tissue. The succeeding involvement of myogenic regulatory factors MyoD, Myf5, Mrf4, and myogenin (MRFs) leads to the fusion of differentiated myoblasts to multinucleated myotubes. Combined into myofibrils, these are packed together to form a muscle fibre (myocyte—m_f).

A number of myoblasts, satellite cells, remain in a quiescent state. These are crucial for muscle regeneration and repair in response to mechanical stimuli (loading, straining) and/or damage. Although at the time of writing little information was available regarding the expression kinetics of IGF-1 in exercise-injured muscle, it has been proposed that the peptide is synthesised and released by satellite cells and by the muscle itself (Chakravarthy *et al* 2000, Jennische *et al* 1987, Hawke and Garry 2001, Adams 2002). Thus *in vitro* experiments on cyclic mechanical stretching of myoblasts within a physiological limit (10%–20% strain), have shown myoblast proliferation as a result of stimulation of the PI3K/PkA and MAPK pathways (Fu *et al* 2018). The cultured scaffolds in human urinary bladder smooth muscle cells show an increase in proliferation in response to periodic hydrodynamic pressure stimulation ranging within 0–300 cm H_2O. Using flow cytometry, PCR and immunoblot methods, Chen *et al* (2012) implicate the PI3K/serum-glucocorticoid regulated kinase 1 molecular mechanism in the reaction (2012). Additionally, IGF-1 promotes functional changes in the excitability of myocytes and neurons by inducing the translocation of voltage-dependent L- and N-type Ca^{2+} channels to the plasma membrane through the PI3K/PkB signalling pathway (Viard *et al* 2004). The chain of events of the myogenesis are summarised in figure 4.9.

The mathematical equivalent to the proposed scheme is given by

$$d\mathbf{X}_M/dt = \mathbf{A}_8\mathbf{X}_M(t) \tag{4.22}$$

where $\mathbf{A}_8$ is the matrix of rate constants and the vector $\mathbf{X}_M(t) = (x_{109-114})^{\mathrm{T}}$ is defined by

$$x_{109} = [Mef2], \quad x_{110} = [Mef2^{\mathrm{A}}], \quad x_{111} = [m_b]$$
$$x_{112} = [m_b^{\mathrm{A}}], \quad x_{113} = [MRFs], \quad x_{114} = [m_f^{sm}].$$

The system (4.22) can be reduced to

$$d\mathbf{X}_M/dt = \mathbf{A}_8\mathbf{X}_M(t) + \mathbf{C}_M \tag{4.23}$$

where $\mathbf{X}_M(t) = (x_{113}, x_{114})^{\mathrm{T}}$, $\mathbf{C}_M = (k_{86}x_{110}x_{111}, 0)^{\mathrm{T}}$ and the reduced matrix $\mathbf{A}_8(2 \times 2)$ has the non-zero elements

$$i_{11} = -i_{21} = -k_{87}x_{112}, \qquad i_{22} = -k_{88}.$$

The substrates $x_{109-112}$ are found from

$$\begin{aligned}
x_{109}(t) &= \frac{k_{86}x_{109}^0x_{111}}{k_{85}x_{108} + k_{86}x_{111}} \\
x_{110}(t) &= \frac{k_{85}x_{109}^0x_{108}}{k_{85}x_{108} + k_{86}x_{111}} \\
x_{111}(t) &= \frac{k_{87}x_{111}^0x_{113}}{k_{86}x_{110} + k_{87}x_{113}} \\
x_{112}(t) &= \frac{k_{86}x_{111}^0x_{110}}{k_{86}x_{110} + k_{87}x_{113}}.
\end{aligned} \tag{4.24}$$

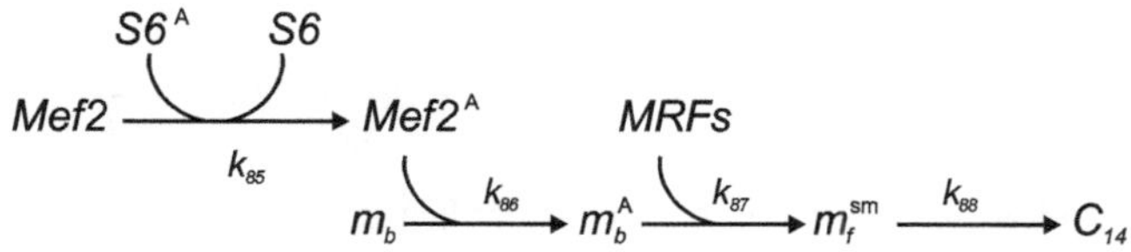

Figure 4.9. Proposed scheme of smooth muscle myogenesis.

Provided the initial concentrations of the substrates, $\mathbf{X}_M(0) = \mathbf{X}_{M0}$, are known, the systems (4.23) and (4.24) describes the myogenesis of smooth muscle.

4.10 Collagenesis

Collagen is a biological polymer with a molecular weight of 300 kDa. This is composed of the four amino acids: glycine, (hydroxy)proline and hydroxylisine arranged in a strictly regular pattern. Collagen synthesis by fibroblasts is controlled by IGF-1 through the PI3K/PkB/S6k-signalling pathway. Thus, activated ribosomal protein S6 kinase (S6k) transcribes genes, COLα1/α3, coded for α1/3 peptide. The expressed mRNA is translated to produce the pre-pro-polypeptide (Pre-pro-α1/α3) which undergoes further post-translational modifications (Blackstock *et al* 2014). These include the signal peptide removal (e_1^1), hydroxylation (e_1^2) and glycosylation (e_1^3) of the lysine and proline residues, the formation of pro-α1/3-chains and their assembly in a left-handed helix—pro-collagen molecule (Pro-c). Packed in secretory vesicles and released into the extracellular space, this is cleaved by procollagen peptidases (e_2) to become tropocollagen (Trop-c). The three helices are covalently cross-linked through an aldol condensation reaction catalysed by lysyl oxidase (e_3) and spontaneously twisted into a right-handed triple helix to form a collagen fibril (C_f), a morphostructural unit of the collagen fibre (figure 4.10).

Molecular pathways involved in the turnover of soft tissue ECM collagen in tissue-remodelling processes include: (i) the membrane-associated matrix metalloproteinases and collagenases (e_4) which directly cleave collagen in the extracellular space, and (ii) the binding of collagen fibrils to specific cell surface receptors, β_1 integrins or urokinase plasminogen activator receptor-associated protein, followed by their cellular uptake and proteolytic degradation by cathepsins (Wagenaar-Miller *et al* 2007).

Applying the principle of mass-action to the proposed scheme, the system of equations of collagenesis yields

$$d\mathbf{X}_C/dt = \mathbf{A}_9\mathbf{X}_C(t) \tag{4.25}$$

where the vector $\mathbf{X}_M(t) = (x_{115-123})^{\mathrm{T}}$ is defined by

$$x_{115} = [COL1\alpha 1/3], \quad x_{116} = [\mathrm{Pre} \cdot pro \cdot \alpha 1/3], \quad x_{117} = [e_1^{1,2,3}]$$
$$x_{118} = [\mathrm{Pro} \cdot C], \quad x_{119} = [e_2], \quad x_{120} = [Trop \cdot C]$$
$$x_{121} = [e_3], \quad x_{122} = [m_f^C], \quad x_{119} = [e_4]$$

and the matrix of rate constants $\mathbf{A}_9(9 \times 9)$ is

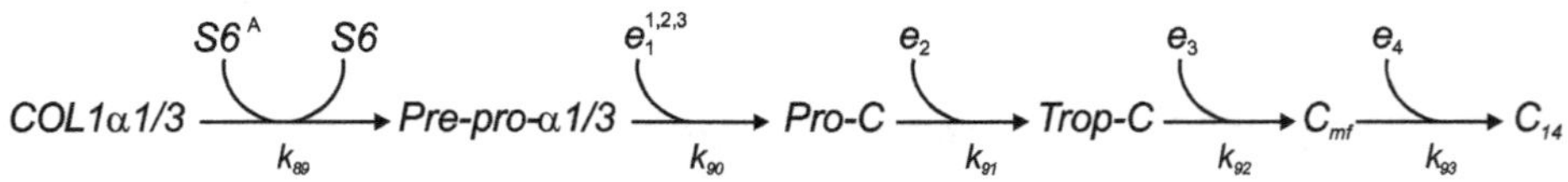

Figure 4.10. Sequence of reactions leading to collagen synthesis.

$$j_{11} = -j_{21} = -k_{89}x_{108}, \quad j_{22} = k_{90}x_{117}, \quad j_{33} = -j_{43} = -k_{90}x_{116}$$
$$j_{44} = -k_{91}x_{119}, \quad j_{55} = -k_{90}x_{118}, \quad j_{44} = k_{91}x_{118}$$
$$j_{66} = -k_{92}x_{121}, \quad j_{77} = -j_{87} = -k_{92}x_{120}, \quad j_{88} = -k_{93}x_{123}$$
$$j_{91} = k_{94}, \quad j_{99} = -k_{93}x_{122}.$$

To close the system the initial concentrations, $\mathbf{X}_C(0) = \mathbf{X}_{C0}$, should be given.

References

Adams G R 2002 Autocrine and/or paracrine insulin-like growth factor-1 activity in skeletal muscle *Clin. Orthop. Relat. Res.* **403** S188–96

Ahmadi S, Muth-Selbach U, Lauterbach A, Lipfert P, Neuhuber W L and Zeilhofer H U 2003 Facilitation of spinal NMDA receptor currents by spillover of synaptically released glycine *Science* **300** 2094–7

Akamizu T *et al* 2004 Pharmacokinetics, safety, and endocrine and appetite effects of ghrelin administration in young healthy subjects *Endocrine* **150** 447–55

Albrecht P, Lewerenz J, Dittmer S, Noak R, Maher P and Methner A 2010 Mechanisms of oxidative glutamate toxicity: the glutamate/cysteine antiporter system xc^- as a neuro-protective drug target *CNS Neur. Disord. Drug. Targets* **9** 373–82

Alford F P, Baker H W G and Burger H G 1973 The secretion rate of human growth hormone. I. Daily secretion rates, effect of posture and sleep *J. Clin. Endocr. Met.* **37** 515–20

Anderson L L, Jeftinija S and Scanes C G 2004 Growth hormone secretion: molecular and cellular mechanisms and *in vivo* approaches *Exp. Biol. Med.* **229** 291–302

Ariyasu H *et al* 2001 Stomach is a major source of circulating ghrelin, and feeding state determines plasma ghrelin-like immunoreactivity levels in *humans J. Clin. Endocrinol. Met.* **86** 4753–8

Arvat E *et al* 2001 Endocrine activities of ghrelin, a natural growth hormone secretagogue (Ghs), in humans: comparison and interactions with hexarelin, a nonnatural peptidyl Ghs, and Gh-releasing hormone *J. Clin. Endocr. Met.* **86** 1169–74

Asa S L, Coschigano K T, Bellush L, Kopchick J J and Ezzar S 2000 Evidence for growth hormone (GH) autoregulation in pituitary somatotrophs in GH antagonist transgenic mice and GH receptor-deficient mice *Am. J. Path.* **156** 1009–15

Aubrey K R and Supplisson S 2018 Heterogeneous signaling at GABA and Glycine co-releasing terminals *Front Synaptic Neurosci.* **10** 40

Baccam N, Alonso G, Costecalde T, Fontanaud P, Molino F, Robinson I C A F, Mollard P and Mery P-F 2007 Dual-level afferent control of growth hormone-releasing hormone (GHRH) neurons in GHRH-green fluorescent protein transgenic mice *J. Neurosci.* **27** 1631–41

Balthazar N, Mery P-F, Magoulas C B, Mathers K E, Martin A, Mollard P and Robinson I C A F 2003 Growth hormone-releasing hormone (GHRH) neurons in GHRH-enhanced green fluorescent protein transgenic mice: a ventral hypothalamic network *Endocrine* **144** 2728–40

Barabitus N and Schally A V 2010 Growth hormone-releasing hormone. Extrapituitary effects in physiology and pathology *Cell Cycle* **9** 4110–6

Barberis A, Mozrzymas J W, Orlinski P I and Vicini S 2007 Desensitization and binding properties determine distinct $\alpha 1\beta 2\gamma 2$ and $\alpha 3\beta 2\gamma 2$ $GABA_A$ receptor-channel kinetic behavior *Eur. J. Neurosci.* **25** 2726–40

Barclay J L, Kerr L M, Arthur L, Rowland J E, Nelson C N, Ishikawa M, d'Aniello E M, White M, Noakes P G and Waters M J 2010 *In vivo* targeting of the growth hormone receptor (GHR) Box1 sequence demonstrates that the GHR does not signal exclusively through JAK2 *Mol. Endocr.* **24** 204–17

Barnett P 2003 Somatostatin and somatostatin receptor physiology *Endocrine* **20** 255–64

Bennett P A, Thomas G B, Howard A D, Feighner S D, van der Ploeg L H, Smith R G and Robinson I C 1997 Hypothalamic growth hormone secretagogue-receptor (GHS-R) expression is regulated by growth hormone in the rat *Endocrine* **138** 4552–7

Bentov I and Werner H 2013 Insulin-like growth factor-1 *Handbook of Biologically Active Peptides* (Amsterdam: Elsevier) ch 222

Bernat B, Pal G, Sun M and Kossiakoff A A 2003 Determination of the energetics governing the regulatory step in growth hormone-induced receptor homo-dimerization *Proc. Natl Acad. Sci. USA* **100** 952–7

Betley J N, Cao Z F, Ritola K D and Sternson S M 2013 Parallel, redundant circuit organization for homeostatic control of feeding behaviour *Cell* **155** 1337–50

Bhat G K, Mahesh V B, Chu Z W, Chorich L P, ZAmorano P L and Brann D W 1995 Localization of N-methyl-D-aspartate R1 receptor subunit in specific anterior pituitary hormone cell types of the female rat *Neuroendocrine* **62** 178–86

Bjelobaba I, Janjic M M and Stojilkovic S S 2015 Purinergic signaling pathways in endocrine system *Auton. Neurosci.* **191** 102–16

Blackstock C, Higashi Y, Sukhanov S, Shai S-Y, Stefanovic B, Tabony A M, Yoshida T and Delafontaine P 2014 Insulin-like growth factor-1 increases synthesis of collagen type I via induction of the mRNA-biding protein LARP6 expression and binding to the 5′ stem-loop of COL1a1 and COL1a2 mRNA *J. Biol. Chem.* **289** 7264–74

Blakely R D and Edwards R H 2012 Vesicular and plasma membrane transporters for neurotransmitters *Gold Spring Harb. Perspect. Biol.* **4** a005595

Bonert C S and Melmed S 2017 *The Pituitary* (Amsterdam: Elsevier)

Breder C D, Yamada Y, Yasuda K, Seino S, Saper C B and Bell G 1992 Differential expression of somatostatin receptor subtypes in brain *J. Neurosci.* **12** 3920–34

Brogio F, Gottero C, Prodam F, Gauna C, Muccioli G, Papotti M, Arbibat T, van der Lely A J and Grogo E 2004 Non-acylated ghrelin counteracts the metabolic but not the neuroendocrine respose to acylated ghrelin in humans *J. Clin. Endocr. Met.* **87** 3783–90

Brooks A J *et al* 2014 Mechanism of activation of protein kinase JAK2 by the growth hormone receptor *Science* **344** 1249783

Campbell J N *et al* 2017 A molecular census of arcuate hypothalamus and median eminence cell types *Nat. Neurosci.* **20** 484–96

Cao P, Maximov A and Südhof T C 2011 Activity-dependent IGF-1 exocytosis is controlled by the Ca^{2+}-sensor synaptotagmin-10 *Cell* **145** 300–11

Chakravarthy M V, Abraha T W, Schwartz R J, Fiorotto M L and Booth F W 2000 Insuline-like growth factor-I extends *in vitro* replicative life span of skeletal muscle satellite cells by enhancing G1/S cell cycle progression via the activation of phosphatidylinositol 3′-kinase/Akt signalling pathway *J. Biol. Chem.* **275** 35942–52

Chen L, Wei T-q, Wang Y, Zhang J, Li H and Wang K-J 2012 Simulated bladder pressure stimulates bladder smooth muscle cell proliferation via the PI3K/SGK1 signaling pathway *J. Urol.* **188** 661–7

Chen S X, Kim A N, Peters A J and Komiyama T 2015 Subtype-specific plasticity of inhibitory circuits in motor cortex during motor learning *Nat. Neurosci.* **18** 1109–15

Cheng H, Yibchok-anun S, Coy D H and Hsu W H 2002 SSTR2 mediates the somatostatin-induced increase in intracellular Ca^{2+} concentration and insulin secretion in the presence of arginine vasopressin in clonal β-cell HIT-T15 *Life Sci.* **71** 927–36

Chia D J 2014 Minireview: mechanisms of growth hormone-mediated gene regulation *Mol. Endocr.* **28** 1012–25

Clemens J D, Lester R A, Tong G, Jahr C E and Westbrook G L 1992 The time course of glutamate in the synaptic cleft *Science* **258** 1498–501

Córdoba-Chacón J, Gahete M D, Culler M D, Castano J P, Kineman R D and Luque R M 2011 Somatostatin dramatically stimulates growth hormone release from primate somatotrophs acting at low doses via somatostatin receptor 5 and cyclic AMP *J. Neuroendocr.* **24** 453–63

Cui T and Schally A V 2018 Growth hormone-releasing hormone (GHRH) and its agonists inhibit hepatic and tumoral secretion of IGF-1 *Oncotarget* **9** 28745–56

Danbolt N C 2001 Glutamate uptake *Prog. Neurobiol.* **65** 1–105

Dehkhoda F, Lee C M M, Medina J and Brooks A J 2018 The growth hormone receptor: mechanism of recetor activation, cell signaling, and phasiollogical aspects *Front. Endocr.* **9** 35

Delfs J R and Dischter M A 1983 Effects of somatostatin on mammalian cortical neurons in culture: physiological actions and unusual dose response characteristics *J. Neurosci.* **3** 1176–88

De-Miguel F F and Nicholls J D 2015 Release of chemical transmitters from cell bodies and dendrites of nerve cells *Phil. Trans. R. Soc.* B **370** 20140181

Devesa J, Almengló C and Devesa P 2016 Multiple effects of growth hormone in the body: is it really the hormone for growth *Clin. Med. Instr: Endocr. Diab.* **9** 47–71

Diamond J S and Jahr C E 1997 Transporters buffer synaptically released glutamate on a submillisecond time scale *J. Neurosci.* **17** 4672–87

Dixon C, Sah P, Lynch J W and Keramidas A 2014 $GABA_A$ receptor α and γ subunits shape synaptic currents via different mechanisms *J. Biol. Chem.* **289** 5399–411

Duan C, Ren H and Gao S 2010 Insulin-like growth factors (IGFs), IGF receptors, and IGF-binding proteins: roles in skeletal muscle growth and differentiation *Gen. Comp. Endocr.* **167** 344–51

Durkin M M, Smith K E, Borden L A, Weinschank R L, Branchek T A and Gustafson E L 1995 Localization of messenger RNA encoding three GABA transporters in rat brain: an *in situ* hybridization study *Brain. Res. Mol. Brain. Res.* **33** 7–21

Eigler T and Ben-Shlomo A 2014 Somatostatin system: molecular mechanisms regulating anterior pituitary hormone *J. Mol. Endocr.* **53** R1–19

Eyigor O, Centers A and Jennes L 2001 Distribution of ionotropic glutamate receptor subunit mRNA in the rat hypothalamus *J. Comp. Neurol.* **434** 101–24

Farrant M and Nusser Z 2005 Variations on inhibitory theme: phasic and tonic activation of $GABA_A$ receptors *Nature. Rev. Neurosci.* **6** 215–29

Farsi Z, Preobraschenski J, van den Bogaart G, Riedel D, Jahn R and Woehler A 2016 Single-vesicle imaging reveals different transport mechanisms between glutamatergic and GABAergic vesicle *Science* **351** 981–4

Fernandez-Pérez L, Flore-Morales A, Guerra B, Diaz-Chico J C and Iglesias-Gato D 2016 Growth hormone receptor signalling pathways and its negative regulation by SOCS2 *Restricted Growth: Clinical, Genetic and Molecular Aspects* (London: IntechOpen) p 1472

Frank S J and Fuchs S Y 2008 Modulation of growth hormone receptor abundance and function: roles of the ubuquiting-proteasome system *Biochim. Biophys. Acta* **1782** 785–94

Frohman L A, Downs T R, Williams T C, Heimer E P, Pan Y C and Felix M 1986 Rapid enzymatic degradation of growth hormone-releasing hormone by plasma *in vitro* and *in vivo* to a biologically inactive product cleaved at the NH2 terminus *J. Clin. Invest.* **78** 906–13

Fu S, Yin L, Lin X, Lu J and Wang X 2018 Effects of cyclic mechanical stretch on the proliferation of L6 myoblasts and its mechanisms: PI3K/Akt and MAPK signal pathways regulated by IGF-1 receptor *Int. J. Mol. Sci.* **29** 1649

Gahete M D, Córdoba-Chacón J, Lin Q, Brüning J C, Kahn C R, Castaño J P, Chrstian H, Luque R M and Kineman R D 2013 Insulin and IGF-1 inhibit GH synthesis and release *in vitro* and *in vivo* by separate mechanisms *Endocrine* **154** 2410–20

Gautam G *et al* 2009 Neural M_3 muscarinic acetylcholine receptors are essential for somatotroph proliferation and normal somatic growth *Proc. Natl Acad. Sci. USA* **106** 6398–403

Gnanapavan S, Kola B, Bustin S A, Morris D G, McGee P, Fairclough P, Bhattacharya S, Carpenter R, Grossman A B and Korbonits M 2002 The tissue distribution of the mRNA of ghrelin and subtypes of its receptor, GHS-R in humans *J. Clin. Endocr. Met.* **87** 2988–91

Guan X M, Yu H, Palyha O C, McKee K K, Feighner S D, Sirinathsinghji D J, Smith R G, Van der Ploeg L H and Howard A D 1997 Distribution of mRNA encoding the growth hormone secretagoue receptor in brain and peripheral tissues *Brain Res. Mol. Brain Res.* **48** 23–9

Guillemin R 2005 Hypothalamic hormones a.k.a. hypothalamic releasing factors *J. Endocr.* **184** 11–28

Hallberg M and Nyberg F 2012 Growth hormone receptors in the brain and their potential as therapeutic targets in central nervous system disorders *Open Endocr. J.* **6** 27–33

Hamper C S, Mitoma H and Manto M 2018 GABA and glutamate: their transmitter role in the CNS and pancreatic isles *GABA And Glutamate: New Developments in Neurotransmission Research* (London: IntechOpen) p 2451

Hataya Y *et al* 2001 A low dose of ghrelin stimulates growth hormone (GH) release synergistically with GH-releasing hormone in humans *J. Clin. Endocr. Met.* **86** 4552–5

Hawke T J and Garry D J 2001 Myogenic satellite cells, physiology to molecular biology *J. Appl. Physiol.* **91** 534–51

Herington A C and Lobie P E 2012 Signal transduction mechanisms underlying growth hormone receptor action *Open Endocr. J.* **6** 13–21

Herman M A, Trimbuch T and Rosenmund C 2018 Differential pH dynamics in synaptic vesicles from intact glutamatergic and GABAergic synapses *Front Synaptic Neurosci.* **10** 44

Hook V, Funkelstein L, Lu D, Bark S, Wegrzyn J and Hwang S R 2008 Proteases for processing proneuropeptides into peptide neurotransmitters and hormones *Ann. Rev. Pharmac. Toxicol.* **48** 393–423

Hori T and Takahashi T 2012 Kinetics of synaptic vesicle refilling with neuro-transmitter glutamate *Neuron.* **76** 511–7

Hrabovszky E, Turi G F and Liposits Z 2005 Presence of vesicular glutamate transporter-2 in hypophysiotropic somatostatin but not growth hormone-releasing hormone neurons of the male rat *Eur. J. Neurosci.* **21** 2120–6

Ibata Y and Watanabe K 1977 A morphological survey of the median eminence: fluorescence histochemistry, electron microscopy and immunohistochemistry *Arch. Hist. Jap.* **40** 303–15

Inenaga K, Dyball R E J, Okuya S and Yashamita H 1986 Characterization of hypothalamic noradrenaline receptors in the supraoptic nucleus and periventricular region on the paraventricular nucleus of mice *in vitro Brain Res.* **369** 37–47

Jennische E, Skottner A and Ha H 1987 Satellite cells express the trophic factor IGF-1 in regenerating skeletal muscle *Acta Physiol. Scand.* **129** 9–15

Jeong J H, Woo Y J, Chua S and Jo Y H 2016 Single-cell gene expression analysis of cholinergic neurons in the arcuate nucleus of the hypothalamus *PLoS One* **11** e0162839

Kakucska I, Tappaz M L, Gaal G, Stoekel M E and Makara G B 1988 Gabaergic innervation of somatostatin-containing neurosecretory cells of the anterior periventricular hypothalamic area: a light and electron microscopy double immunolabeling study *Neuroscience* **25** 585–93

Kanner B I 1994 Sodium-coupled neurotransmitter transport: structure, function and regulation *J. Exp. Biol.* **196** 237–49

Kavran J M *et al* 2014 How IGF-1 activates its receptor *Elife* **25** e03772

Kim H-S, Rosenfeld R G and Oh Y 1997 Biological roles of insulin-like growth factor binding proteins (IGFBPs) *Exp. Mol. Med.* **29** 85–96

Kineman R D, Rio-Moreno M and Sarmento-Cabral A 2018 40 years of IGF-1 understanding the tissue-specific roles of IGF1/IGF1R in regulating metabolism using the Cre/loxP system *J. Mol. Endocr.* **61** T187–98s

Kiss J, Csaba Z, Csáki A and Halász B 2006 Glutamatergic innervation of growth hormone-releasing-containing neurons in the hypothalamic arcuate nucleus and somatostatin-containing neurons in the anterior periventricular nucleus of the rat *Brain Res. Bull.* **70** 278–88

Kiss J, Kocsis K, Csaki Á, Görcs T J and Halász B 1997 Metabotropic glutamate receptor in GHRH and b-endorphin neurones of the hypothalamic arcuate nucleus *Neuro. Rep.* **8** 3703–7

Korbonits M, Bustin S A, Kojima M, Jordan S, Adams E F, Lowe D G, Kangawa K and Grossman A B 2001b The expression of the growth hormone secretagogue receptor ligand ghrelin in normal and abnormal human pituitary and other neuroendocrine tumors *J. Clin. Endocr. Met.* **86** 881–7

Korbonits M, Kojima M, Kangawa K and Grossman A B 2001a Presence of ghrelin in normal and adenomatous human pituitary *Endocrine* **14** 101–4

Kreutzberger A J B, Kiessling V, Liang B, Seelheim P, Jakhanwal S, Jahn R, Castle D and Tamm L K 2017 Reconstruction of calcium-mediated exocytosis of dense-core vesicles *Sci. Adv.* **3** e1603208

Kumar U 2012 Immunohistochemical distribution of somatostatin and somatostatin receptor subtypes (sst1–5) in hypothalamus of ApoD knockout mice brain *J. Mol. Neurosci.* **48** 684–95

Kumar U 2013 G-protein coupled receptors dimerization: diversity in somatostatin receptors subtypes *J. Pharmacogen. Pharcoprot.* **4** 2

Kurosumi K, Koyama T and Tosaka H 1986 Three types of growth hormone cells of the rat anterior pituitary as revealed by immunoelectron microscopy using a colloidal gold-antibody method *Arch. Hist. Jap.* **49** 227–42

Labrakakis C, Rudolph U and Koninck Y D 2014 The heterogeneity in $GABA_A$ receptor-mediated IPSC kinetics reflects heterogeneity of subunit composition among inhibitory and excitatory interneurons in spinal lamina II *Front Cell Neurosci.* **8** 424

Lahlou H, Guilleermet J, Hortala M, Vernejoul F, Pyronnet S, Bousquet C and Susini C 2004 Molecular signalling of somatostatin receptors *Ann. N.Y. Acad. Sci.* **1014** 121–31

Lanneau C, Peineau S, Petit F, Epelbaum J and Gardette R 2000 Somatostatin modulation of excitatory transmission between periventricular and arcuate hypothalamic nuclei *in vitro Am. J. Phys.* **84** 1464–74

Lanneau C, Viollet C, Faivre-Bauman A, Loudes C, Kordon C, Epelbaum J and Gardette R 1998 Somatostatin receptor subtypes sstl and sst2 elicit opposite effects on the response to glutamate of mouse hypothalamic neurons: and electrophysiological and single cell RT-PCR study *Eur. J. Neurosci.* **10** 204–12

Laursen L S, Kjaer-Sorensen K, Andersen M H and Oxvig C 2007 Regulation of insulin-like growth factor (IGF) bioactivity by sequential proteolytic cleavage of IGF binding protein-4 and −5 *Mol. Endocr.* **21** 1245–57

Laviano A, Molffino A, Rianda S and Fanelli F R 2012 The growth hormone secretagogue receptor (Ghs-R) *Curr. Pharmaceut. Design* **18** 4749–54

Lee J-S, Mayers M S, Stromer M H, Scanes C G, Jeftina S and Andersson L A 2004 Number of secretory vesicles in growth hormone cells of the pituitary remains unchanged after secretion *Exp. Biol. Med.* **229** 632–9

Lelouvier B *et al* 2008 Dynamics of somatostatin type 2A receptor cargoes in living hippocampal neurons *J. Neurosci.* **28** 4336–9

Levy L, Bourdais J, Mouhieddine B, Benlot C, Villares S, Cohen P, Peillon F and Joubert D 1993 Presence and characterization of the somatostatin precursor in normal human pituitaries and in growth hormone secreting adenomas *J. Clin. Endocr. Met.* **76** 85–90

Liguz-Lecznar M, Urban-Ciecko J and Kossut M 2016 Somatostatin and somatostatin-containing neurons in shaping neuronal activity and plasticity *Front Neur. Circ.* **10** 48

Linossi E M, Babon J J, Hilton D J and Nicholson S E 2013 Suppression of cytokinesignaling: the SOCS perspective *Cytokine Growth Factor Rev.* **24** 241–8

Liu Q, Cescato R, Dewi D A, Rivier J, Reubi J C and Schonbrunn A 2005 Receptor signalling and endocatosis are differentially regulated by somatostatin analogs *Mol. Pharm.* **68** 90–101

Lobie P E, Garcla-Aragón J, Lincoln D T, Barnard R, Wilcox J N and Waters M J 1993 Localization and ontogeny of growth hormone receptor gene expression in the central nervous system *Dev. Brain Res.* **74** 225–33

Lussier B T, French M B, Moor B C and Kraicer J 1991a Free intracellular Ca^{2+} concentration ($[Ca^{2+}]_i$) and growth hormone release from purified rat somatotrophs. I. GH-releasing fctor-induced Ca^{2+} influx raises $[Ca^{2+}]_i$ *Endocrine* **128** 570–82

Lussier B T, French M B, Moor B C and Kraicer J 1991b Free intracellular Ca^{2+} concentration ($[Ca^{2+}]_i$) and growth hormone release from purified rat somatotrophs. III. Mechanism of action of GH-releasing factor and somatostatin *Endocrine* **128** 592–603

Lussier B T, Wood D A, French M B, Moor B C and Kraicer J 1991c Free intracellular Ca^{2+} concentration ($[Ca^{2+}]_i$) and growth hormone release from purified rat somatotrophs. II. Somatostatin lowers $[Ca^{2+}]_i$ by inhibiting Ca^{2+} influx *Endocrine* **128** 583–91

Mancillas J R, Siggins G R and Bloom F E 1986 Somatostatin selectively enhances acetylcholine-induced excitations in rat hippocampus and cortex *Proc. Natl. Acad. Sci. U.S.A.* **83** 7518–21

Mano-Otagiri A, Nemoto T, Sekino A, Yamauchi N, Shuto Y, Sugihara H, Oikawa S and Shibasaki T 2006 Growth hormone-releasing hormone (GHRH) neurons in the arcuate nucleus (Arc) of the hypothalamus are decreased in transgenic rats whose expression of ghrelin receptor is attenuated: evidence that ghrelin receptor is involved in the up-regulation of GHRH expression in the Arc *Endocrine* **147** 4093–103

Martinez V 2013 Somatostatin *Handbook of Biologically Active Peptides* (Amsterdam: Elsevier)

Matsoukas M T and Spyroulias G A 2017 Dynamic properties of the growth hormone releasing hormone receptor (GHRHR) and molecular determinants of GHRH binding *Mol. Biosyst.* **13** 1313–22

Matthews C C, Zielke H R, Wollack J B and Fishman P S 2000 Enzymatic degradation protects neurons from glutamate excitotoxicity *J. Neurochem.* **75** 1045–52

Mayer M L and Armstrong N 2004 Structure and function of glutamate receptor ion channels *Ann. Rev. Physiol.* **66** 161–81

Mayo K E, Miller T, DeAlmeida V, Godfrey P, Zheng J and Cunha S 2000 Regulation of the pituitary somatotroph call by GHRH and its receptor *Recent Prog. Horm. Res.* **55** 237–66

Minami S, Kamegai J, Sugihara H, Suzuki N and Wakabayashi I 1998 Growth hormone inhibits its own secretion by acting on the hypothalamus through its receptors on neuropeptide Y neurons in the arcuate nucleus and somatostatin neurons in the periventricular nucleus *Endocr. J.* **45** S19–26

Minelli A, DeBiasi S, Brecha N C, Zuccarello L V and Conti F 1996 GAT-3, a high-affinity GABA plasma membrane transporter, is localized to astrocytic processes, and it is not confined to the vicinity of GABAergic synapses in the cerebral cortex *J. Neurosci.* **16** 6255–64

Nillni E A, Steinmetz R and Pescovitz O H 1999 Posttranslational processing of progrowth hormone releasing hormone *Endocrine* **140** 5817–27

Nowak L, Bregestovski P, Ascher P, Herbert A and Prochiantz A 1984 Magnesium gates glutmate-activated channels in mouse central neurones *Nature* **307** 462–5

Osterstock G *et al* 2016 Somatostatin triggers rhythmic electrical firing in hypothalamic GHRH neurons *Sci. Rep.* **6** 24394

Pang Z P and Südhof T C 2010 Cell biology of Ca^{2+} triggered exocytosis *Curr. Opin. Cell Biol.* **22** 496–505

Pardridge W M, Eisenberg J S and Yamada T 1985 Rapid sequestration and degradation of somatostatin analogues by isolated brain microvessels *J. Neurochem.* **44** 1178–84

Patel Y C 1999 Somatostatin and its receptor family *Front Neuroendocr.* **20** 157–98

Petralia R S and Wenthold R J 1992 Light and electron immunohistochemical localization of AMPA-selective glutamate receptors in the rat brain *J. Comp. Neurol.* **318** 329–54

Pilecka I *et al* 2007 Protein tyrosine phosphatase H1 (PTP-H1/FTPN3) controls growth hormone receptor signalling and systemic growth *J. Biol. Chem.* **282** 35405–15

Pin J P and Acher F 2002 The metabotropic glutamate receptors: structure, activation mechanism and pharmacology *Curr. Drug Targets CNS Neurol. Disord.* **1** 297–317

Plotzky P M and Vale W 1985 Patterns of growth hormone-releasing factor and somatostatin secretion into the hypophysial portal circulation of the rat *Science* **230** 461–3

Pombo C M, Zalvide J, Gaylinn B D and Dieguez C 2000 Growth hormone-releasing hormone stimulates mitogen-activated protein kinase *Endocrine* **141** 2113–9

Radian R, Bendahan A and Kanner B I 1986 Purification and identification of the functional sodium- and chloride-coupled gamma-aminobutyric acid transport glycoprotein from rat brain *J. Biol. Chem.* **261** 15437–41

Radian R, Ottersen O P, Storm-Mathisen J, Castel M and Kanner B I 1990 Immunocytochemical localization of the GABA transporter in rat brain *J. Neurosci.* **10** 1319–30

Rawlings S R and Hezareh M 1996 Pituitary adenylate cyclase-activating polypeptide (PACAP) and PACAP/vasoactive intestinal polypeptide receptors: action on the anterior pituitary gland *Endocr. Rev.* **17** 4–29

Reichenbach A, Steyn F J, Sleeman M W and Andrews Z B 2012 Ghrelin receptor expression and colocalization with anterior pituitary hormones using a GHSR-GFP mouse line *Endocrin.* **153** 5452–66

Rocheville M, Lange D C, Kumar U, Sasi R, Patel R C and Patel Y C 2000 Subtypes of the somatostatin receptor assemble as functional homo- and heterodimers *J. Biol. Chem.* **275** 7862–9

Romero C J, Pine-Twaddell E, Sima D I, Miller R S, He L, Wondisford F and Radovick S 2012 Insulin-like growth factor 1 mediates negative feedback to somatotroph GH expression via POU1F1/CREB binding protein interactions *Mol. Cell. Biol.* **32** 4258–69

Roth F C and Draguhn A 2012 GABA metabolism and transport: effects on synaptic plasticity *Neural Plast.* **2012** 805830

Rowley N M, Madsen K K, Schouboe A and White H S 2012 Glutamate and GABA synthesis, release, transport and metabolism as targets for seizure control *Neurochem. Int.* **61** 546–58

Rowlinson S W *et al* 2008 An agonist-induced conformational change in the growth hormone receptor determines the choice of signalling pathway *Nat. Cell Biol.* **10** 740–7

Russo A F 2017 Overview of neuropeptides: awakening the senses? *Headache* **57** 37–46

Sasaki F 1987 Changes with age in the number and size of anterior pituitary cells in female mice from suckling to adulthood *J. Endocr.* **117** 5–10

Sato T, Nakamura Y, Shiimura Y, Ohgusu H, Kangawa K and Kojima M 2012 Structure, regulation and function of ghrelin *J. Biochem.* **151** 119–28

Savigny P, Evans J and McGrath K M 2007 Cell membrane structure during exocytosis *Endocrinology* **148** 3863–74

Schaffer F, Baumann G, Haffnenr D, Faunt L M, Johnson M L, Mercado M, Ritz E, Mehls O and Veldhuis J D 1996 Multifactorial control of the elimination kinetics of unbound (free) growth hormone (GH) in the human: regulation by age, adiposity, renal function, and steady state concentrations of GH in plasma *J. Clin. Endocr. Met.* **81** 22–31

Schilbach K and Bidlingmaier M 2015 Growth hormone binding protein—physiological aspects *Best Pract. Res. Clin. Endocr. Met.* **29** 671–83

Schmidt W and Moore H-P H 1994 Synthesis and targeting of insulin-like growth factor-I to the hormone storage granules in an endocrine cell line *J. Biol. Chem.* **269** 27115–24

Schonbrunn A 2009 Selective agonism in somatostatin receptor signalling and regulation *Mol. Cell Endocr.* **14** 35–9

Scimemi A 2014 Structure, function, and plasticity of GABA transorters *Front Cell Neurosci.*

Shigeri Y, Seal R P and Shimamoto K 2004 Molecular pharmacology of glutamate transporters, EAATs and VGLUTs *Brain Res. Rev.* **45** 250–65

Siegel E and Steinmann M E 2012 Structure, functions, and modulation of GABA(A) receptors *J. Biol. Chem.* **287** 40224–31

Stojilkovic S S, Tabak J and Bertram R 2010 Ion channels and signaling in the pituitary gland *Endocr. Rev.* **31** 845–915

Stroh T, van Schouwenburg M R, Beaudet A and Tannenbaum G S 2009 Subcellular dynamics of somatostatin receptor subtype 1 in the rat arcuate nucleus: receptor localization and synaptic connectivity vary in parallel with the ultradian rhythm of growth hormone secretion *J. Neurosci.* **29** 8198–205

Südhof T C 2013 A molecular machine for neurotransmitter release: synapto-tagmin and beyond *Nat. Med.* **19** 1227–31

Sugihara H, Emoto N, Tamura H, Kamegai J, Shibasaki T, Minami S and Wakabayashi I 1999 Effect of insulin like growth factor-1 on growth hormone-releasing factor receptor expression in primary rat anterior pituitary cell culture *Neurosci. Lett.* **276** 87–90

Sun L and Miller R J 1999 Multiple neuropeptide Y receptors regulate K^+ and Ca^{2+} channels in acutely isolated neurons from the rat arcuate nucleus *J. Neurophysiol.* **81** 1391–403

Takahashi T, Forsythe I D, Tsujimoto T, Barnes-Davies M and Onodera K 1996 Presynaptic calcium current modulation by a metabotropic glutamate receptor *Science* **274** 594–7

Tannenbaum G C and Ling N 1984 The interrelationship of growth hormone (GH)-releasing factor and somatostatin in the generation of the ultradian rhythm of GH secretion *Endocrine* **115** 1952–7

Tannenbaum G S, Zhang W H, Lapointe M, Zeitler P and Beaduet A 1998 Growth hormone-releasing hormone neurons in the arcuate nucleus express both sstl and sst2 somatostatin receptor genes *Endocrine* **139** 1450–3

Tedfort H W and Zamponi G W 2006 Direct G protein modulation of Cav2 calcium channels *Pharm. Rev.* **58** 837–62

Tomić M, Zidanovich D, Van Goor F, Yuan D, Koshikizu T and Stojilkovic S S 1999 expression of Ca^{2+}-mobilizing endothelin A receptors and their role in the control Ca^{2+} influx and growth hormone secretion in pituitary somatroprophs *J. Neurosci.* **19** 7721–31

Traynelis S, Hartley M and Heinemann S F 1995 Control of proton sensitivity of the NMDA receptor by RNA splicing and polyamines *Science* **268** 873–6

van den Pol A, Hermans-Borgmeyer I, Hofer M, Ghosh P and Heinemann S 1994 Ionotropic glutamate-receptor gene expression in hypothalamus: localization of AMPA, Kainate, and NMDA receptor RNA with *in situ* hybridization *J. Comp. Neurol.* **343** 428–44

Viard P, Butcher A J, Halet G, Davies A, Nürnberg B, Heblich F and Dolphin A C 2004 PI3K promotes voltage-dependent calcium channel trafficking to the plasma membrane *Nature Neurosci.* **7** 939–46

Vijayakumar A, YAkar S and LeRoith D 2011 The intricate role of growth hormone in metabolism *Front Endocr.* **2** 32

Vilardaga J-P, Agnati L F, Fuxe K and Ciruela F 2010 G-protein-coupled heteromer dynmaics *J. Cell Sci.* **123** 4215–20

Wagenaar-Miller R A, Engelholm L H, Gavard J, Yamada S S, Gutkind J S, Behrendt N, Bugge T H and Holmbeck K 2007 Complementary roles of intracellular and pericellular collagen degradation pathways *in vivo Mol. Cell. Biol.* **27** 6309–22

Waters M J 2015 The growth hormone receptor *Growth Horm. IGF Res.* **28** 6–10

Waters M J, Brooks A J and Chabra Y 2014 A new mechanism for growth hormone receptor activation of JAK2, and implications for related cytokine receptors *JAKSTAT* **3** e29569

Weckbecker G, Lewis I, Albert R, Schmid H A, Hoyer D and Bruns C 2003 Opportunities in somatostatin research: biological, chemical and therapeutic aspects *Nature Rev. Drug Dis.* **2** 999–1017

Werther G A, Hogg A, Oldfield B J, McKinley M J, Figdor R and Mendelsohn F A O 1989 Localization and characterization of insulin-like growth factor-1 receptors in rat brain and pituitary gland using *in vitro* autoradiography and computerized densitometry* a distinct distribution from insulin receptors *J. Neuroendocr.* **1** 369–77

Winer L M, Shaw M A and Baumann G 1990 Basal plasma growth hormone levels in man: new evidence for rhythmicity of growth hormone secretion *J. Clin. Endocr. Met.* **70** 1678–86

Xu Y, Kong G K-W, Menting J G, Margetts M B, Delaine C A, Jenkin L N, Kiselyov V V, De Meyts P, Be F and Lawrence M C 2018 How ligand binds to the type 1 insulin-like growth factor receptor *Nature. Comm.* **9** 821

Yamashita M, Kawaguchi S Y, Hori T and Takahashi T 2018 Vesicular GABA uptake can be rate limiting for recovery of IPSCs from synaptic depression *Cell Rep.* **22** 3134–41

Yang J, Brown M S, Liang G, Grishin N V and Goldstein J L 2008 Identification of the acyltransferase that octanoylates ghrelin, an appetite-stimulating peptide hormone *Cell* **132** 387–96

Yang S K and Chen C 2007 Involvement of somatostatin receptor subtypes in membrane ion channel modification by somatostatin in pituitary somatotropes *Clin. Exp. Pharmac. Physiol.* **34** 1221–7

Yin Y, Li Y and Zhang W 2014 The growth hormone secretagogue receptor: its intracellular signalling and regulation *Int. J. Mol. Sci.* **15** 4837–55

Zemkova H and Stojilkovic 2018 Neurotrasmitter receptors as signalling platforms in anterior pituitary cells *Mol. Cell Endocr.* **463** 49–64

Zhang X and van den Pol 2015 Dopamine/tyrosine hydroxylase neurons of the hypothalamic arcuate nucleus release GABA, communicate with dopaminergic and other arcuate neurons, and respond to dynorphin, met-enkephalin, and oxytocin *J. Neurosci.* **35** 14966–82

Zheng H, Bailey A, Jiang M-H, Honda K, Chen H Y, Trumbauer M E, van der Ploeg L H T, Schaeffer J S, Leng G and Smith R G 1997 Somatostatin receptor subtype 2 knockout mice are refractory to growth hormone-negative feedback on arcuate neurons *Mol. Endocr.* **11** 1709–17

Zigman J M, Jones J E, Lee C E, Saper C B and Elmquist J K 2006 Expresssion of ghrelin receptor mRNA in the rat and in the mouse brain *J. Comp. Neurol.* **494** 528–48

Chapter 5

Modelling the (intra)hypothalamic–pituitary axis

Common sense is genius dressed in its working clothes.

— Ralph W Emerson

Existing mathematical models of electrical impulse transduction along nerve fibres and somas of cells, protein-tyrosine receptor intracellular signalling pathways, hormonal interactions and gene expression are presented. Modern trends and their pitfalls in the various modelling approaches are discussed.

5.1 Signal transduction

The processes of signal transduction, transcription and translation comprise the fundamental sequence of interlinked extracellular and intracellular events. The basis of cell signalling is the transmission of extracellular electrical and chemical stimuli into the cell to ensure an appropriate physiological response, i.e. cell growth, proliferation, metabolism, remodelling. Electrical activity is ubiquitous and essential in maintaining the vital functions of the body. It manifests in a variety of discharge patterns (action potentials, spikes) and the specificity of transduction mechanisms enable the system to integrate and coordinate the overall behaviours in space and time.

When considering the mathematical modelling of neuronal electrical activity two types of complexity must be dealt with: the interplay of ion channel dynamics that underlie excitability, and neuronal morphology, i.e. dendrites, axons, synapses, that allow neurons to communicate signals among them. Despite significant advancements in simulation and the large number of mathematical models of excitable media proposed, the only existing biologically plausible model is the

doi:10.1088/978-0-7503-3567-6ch5

Hodgkin–Huxley (H–H) model (1952). It contains a current balance equation that satisfies Kirchhoff's law

$$C\frac{dV}{dt} = -\sum_{i=1}^{n} g_i x_i^p y_i^p (V - V_i) + I_{\text{appl}}(t), \tag{5.1}$$

and relaxation equations for the ionic conductances

$$\begin{aligned} \frac{dx_i}{dt} &= \frac{x_{i\infty}(V) - x_i}{\tau_{x_i}(V)}, \quad i = \overline{1, n}, \\ \frac{dy_i}{dt} &= \frac{y_{i\infty}(V) - y_i}{\tau_{y_i}(V)}, \quad i = \overline{1, n}. \end{aligned} \tag{5.2}$$

Here C is the membrane capacitance, V is the membrane potential, V_i is the Nernst potential for the ith ion, g_i is the maximal conductance of the channel for the different ion-selective channels, x_i^p ($x_{i\infty}$), $y_i^p(y_{i\infty})$ are the actual (steady) state activation and inactivation variables, respectively, τ_{x_i} and τ_{y_i} are the relaxation time constants, I_{appl} is the applied current. The nonlinear system (5.1), (5.2) accurately reproduces the rich dynamic behaviour in various excitable tissues ranging from tonic bursting to chaos, and from stationary to travelling wave phenomena. Such versatility and reliability are the result of mathematical formulation incorporating the explicitly/implicitly intrinsic properties of the cell membrane.

Since then innumerable models of single and large populations of neurons and neuronal fields have been formulated and demonstrate their ability to describe the membrane potential and ion current dynamics. However, all of these models are either simplifications or replicas of the H–H model and suffer from constructive biological insufficiencies.

A distinctive feature of neurons is their morphology. Much recent research has been directed at understanding the impact of anatomical architecture on signal integration within a neuron and neuronal ensembles. Originally, a single, electrically equivalent cable model was used to simulate the generation and propagation of voltage waves and current changes in the unmyelinated axon and dendrites (Hodgkin and Huxley 1952)

$$C\frac{\partial V}{\partial t} = \frac{d_{\text{f}}}{2R_{\text{m}}}\frac{\partial^2 V}{\partial x^2} - \sum_{i=1}^{n} g_i x_i^p y_i^p (V - V_i) + I_{\text{appl}}(t), \tag{5.3}$$

where R_{m} is the membrane resistance, d_{f} is the cross-sectional diameter of the nerve fibre, x is the Lagrange coordinate, and the meaning of other parameters and constants is as described above.

As an extension, compartmental models have been derived to simulate and explore the role of cytomorphology, embracing biological mechanisms at molecular and cellular levels to give support to the temporal and spatial propagation and stability of signalling in diverse neuronal topologies, e.g. myelinated fibres, branching structures and synapsing (Zhou and Chiu 2001, Schierwagen and Ohme 2008, Masoli *et al* 2017). Fitzhugh (1962) formulated a mathematical model of the electrical properties of a myelinated nerve fibre consisting of the nodes of Ranvier separated by internodal segments insulated by myelin sheaths. The membrane potential evolution at the nodes follows the general physiological principles of Na^+, K^+ and Cl^- ion channel dynamics and is thus described by

$$C_N \frac{dV_N}{dt} = I_N - A\sum_{i=1}^{n} g_i x_i^p y_i^p (V_N - V_i), \tag{5.4}$$

where the subscript N is referred to the node, C_N is the nodal membrane capacitance, I_N is the nodal outward membrane current, A is the surface area of a node (free of myelin), V_i are the reversal potentials for $i = Na^+$, K^+ and Cl^- currents, respectively.

The propagation of the transmembrane potential, V_{MS}, across the myelin layer in the internodes is given by

$$C_{MS} \frac{\partial V_{MS}}{\partial t} = \frac{1}{R_{MS}} \frac{\partial^2 V}{\partial x^2} - V/r_{MS}, \tag{5.5}$$

where C_{MS} is the myelin capacitance, R_{MS}, r_{MS} are the total longitudinal and myelin resistance, respectively. The continuity of V_{MS} at the nodes is achieved by letting

$$\lim_{x \to j, L} V_{MS}(x, t) = V_{N,j}(t), \quad j = 0, 1, 2, \ldots. \tag{5.6}$$

Here x is the Lagrange coordinate and L the length of the axon, j indicates the position of the node ($x = 2j$ mm) and the internodal space ($2(j - 1)$ mm $< x < 2j$ mm). Boundary conditions at the first node ($x = 0$) and the subsequent nodes ($j = 1, 2, \ldots, L$) provide outward membrane currents

$$\begin{aligned} I_0 &= \left(rI_s + 2\lim_{x \to 0} \partial V_{MS}/\partial x \right) / R_{MS} \\ I_0 &= \left(\lim_{x \to j, L+0} \partial V_{MS}/\partial x - \lim_{x \to j, L-0} \partial V_{MS}/\partial x \right) / R_{MS}. \end{aligned} \tag{5.7}$$

Here I_s is the given stimulating current, and r is the specific input resistance. It is assumed that initially the nerve fibre is in a uniform resting state. Equations (5.4)–(5.7) solved together offer an insight into the 'saltatory' conduction of the wave of depolarisation.

The linear cable theory has been applied to model a continuous composite structure, the myelinated axon, consisting of identical repeating units. Each unit comprises two subunits, one (being) covered with myelin sheath having a length of

L_{MS}, and another being the unmyelinated area, δ_N wide, histomorphologically corresponding to the node of Ranvier. The subunits have different electrical and geometric characteristics (Jack *et al* 1975, Andrietti and Bernardini 1984, Basser 1983). The propagation of V_{MS} is given by a single 'cable' equation

$$\frac{\partial^2 V_{MS}}{\partial x^2} = (1 - \chi(x))\left(\frac{\tau_{MS}}{\lambda_{MS}}\frac{\partial V_{MS}}{\partial t} + \frac{V_{MS}}{\lambda_{MS}^2}\right) + \chi(x)\left(\frac{\tau_N}{\lambda_N}\frac{\partial V_{MS}}{\partial t} + \frac{V_{MS}}{\lambda_N^2} + I_{\text{ion}}(V_{MS}, t)\right). \tag{5.8}$$

The first term describes the transmembrane potential dynamics in the myelinated zone and the second in the node of Ranvier. In the above, $\tau_{MS(N)}$, $\lambda_{MS(N)}$ are the myelin, node time and space constants, whilst $\chi(x)$ is an indicator function represented by superimposing unit Heaviside functions $H(x)$ (Burridge and Keller 1981)

$$\chi(x) = \sum_{n=-\infty}^{\infty}\left(H\left(x - nL_{MS} + \frac{\delta_N}{2}\right) - H\left(x - nL_{MS} - \frac{\delta_N}{2}\right)\right) \tag{5.9}$$

such that $\chi(x) = 1$ if x lies within a node of Ranvier, and 0 otherwise. The axon is initially at a state of physiological rest, $V_{MS} = 0$, and is excited by the electrical current, I_s, injected at the proximal end ($V_{MS,0} = rI_s$).

Since 'saltatory' signal conduction has advantages, i.e. high speed and low energetic requirements, when compared to the incessant propagation along unmyelinated axons. This has led to a growing interest in the study of the intracellular mechanisms involved. For example, Binczak *et al* (2000) have investigated numerically ephaptic (nonsynaptic) interactions and their role in impulse dynamics in a bundle of myelinated nerve fibres. The consequences of subtle paranodal injury as a result of crush or stretch injury on action potential propagation, i.e. retraction and/or detachment, thickening/thinning of myelin, widening of the nodal gap length, and the reaction of astrocytes, has been analysed by Brill *et al* (1977), Zhou and Chiu (2001), Babbs and Shi (2013), Freeman *et al* (2016), Dutta *et al* (2018). Nfor and Mokoli (2016) have studied the effect of dissipation along the myelinated axon on the shape of the travelling signal both analytically and numerically. The interested reader is advised to consult particular publications on the subject which are beyond the scope of this book.

Currently, the only existing mathematical description of electrical activity in the somatotroph is a surrogate model by Tsaneva-Atanasova *et al* (2007). This is based on the Hodgkin–Huxley formalism intergrating L- and T-type Ca^{2+}, delayed and inward-rectifier K^+, non-selective Na^+, two populations of large conductance Ca^{2+}–K^+ (BK_{Ca}) ion and the stochastic noise currents. Some of these were assessed experimentally on cultured rat somatotrophs whilst others were explored in studies on cardiomyocytes, pancreatic beta-cells or adjusted during numerical simulations. Additionally, two intracellular Ca^{2+} compartments, replicas of De Young and

Keizer (1992) and Li and Rinzel (1994) descriptions, were adopted to follow the dynamics of cytosolic Ca^{2+} turnover (appendix C). The governing equation of electrical activity, V_{St}, yields

$$C_{St}\frac{dV_{St}}{dt} = -(I_{CaL} + I_{CaT} + I_{NS,\,Na} + I_{Kdr} + I_{Kir} + I_{BK(n)} + I_{BK(f)}) + I_{Noise}, \quad (5.10)$$

where C_{St} is the specific cell membrane capacitance, and I_{CaL}, I_{CaT} are the L- and T-type Ca^{2+}, I_{Kdr}, I_{Kir} are the delayed-rectifier and inward-rectifier, $I_{NS,Na}$ is the non-selective cAMP regulated Na^+, $I_{BK(n)}$, $I_{BK(f)}$ are the Ca^{2+}–K^+ ion currents, given by

$$\begin{aligned}
I_{CaL} &= g_{CaL}m_{CaT\infty}^2(V_{St} - V_{Ca})\\
I_{CaT} &= g_{CaT}m_{CaT\infty}^2 h_{CaT\infty}(V_{St} - V_{Ca})\\
I_{Kdr} &= g_{Kdr}n_{Kdr}(V_{St} - V_K)\\
I_{Kir} &= g_{Kir}K_{ir\infty}(V_{St} - V_K)\\
I_{BK(n)} &= g_{BK(n)}b_{BK(n)}(V_{St} - V_K)\\
I_{BK(f)} &= g_{BK(n)}b_{BK(f)\infty}(V_{St} - V_K)\\
I_{NS,Na} &= g_{NS,Na}(V_{St} - V_{Na}).
\end{aligned} \quad (5.11)$$

The steady-state activation, $m_{CaL\infty}$, $m_{CaT\infty}$, $K_{ir\infty}$, $b_{BK(n)}$, $b_{BK(f)\infty}$, n_{Kdr}, and inactivation, $h_{CaT\infty}$, functions for the respective ion currents yield

$$\begin{aligned}
m_{CaL\infty} &= 1/[1 + \exp((V_{CaL} - V_{St})/k_{CaL})]\\
m_{CaT\infty} &= 1/[1 + \exp((V_{CaT} - V_{St})/k_{CaT})]\\
h_{CaT\infty} &= 1/[1 + \exp((V_{Cah} - V_{St})/k_{Cah})]\\
K_{ir\infty} &= \alpha_{ir}/(\alpha_{ir} + \beta_{ir})\\
b_{BK(n)} &= 1/\left[1 + \exp\left(\left(V_{BK(n)} - V_{St}\right)/k_{BK}\right)\right]\\
b_{BK(f)} &= 1/\left[1 + \exp\left(\left(V_{BK(f)} - V_{St}\right)/k_{BK}\right)\right]\\
\frac{dn_{Kdr}}{dt} &= \frac{n_{Kdr\infty} - n_{Kdr}}{\tau_{Kdr}}
\end{aligned} \quad (5.12)$$

where

$$\begin{aligned}
n_{Kdr\infty} &= 1/[1 + \exp((V_{Kdr} - V_{St})/k_{Kdr})]\\
\alpha_{ir} &= 0.1/[1 + \exp 0.06(V_{St} - V_{Kdr} - 200)]\\
\beta_{ir} &= \frac{3\exp 0.0002(V_{St} - V_{Kdr} - 100) + \exp 0.1(V_{St} - V_{Kdr} - 10)}{1 + \exp 0.5(V_{Kdr} - V_{St})}\\
V_{BK(n)} &= V_{BK0} - k_{shift}\ln\frac{-0.11(I_{CaL} + I_{CaLT})}{k_{Ca(BK)}}\\
V_{BK(f)} &= V_{BK0} - k_{shift}\ln([Ca^{2+}]_{cyt}/k_{Ca(BK)}).
\end{aligned} \quad (5.13)$$

In the above, V_{BK0}, $V_{K\text{dr}}$, $V_{BK(n)}$, $V_{BK(f)}$, $V_{Ca\text{L}}$, $V_{Ca\text{T}}$, V_{Cah} are the respective reversal potentials, k_{shift}, $k_{Ca\text{T}}$, $k_{Ca\text{L}}$, k_{Cah}, k_{BK}, $k_{K\text{dr}}$, $k_{Ca(BK)}$ are empirical constants, and the dynamics of intracellular calcium turnover, $[\text{Ca}^{2+}]_{cyt}$, given by

$$\begin{aligned}\frac{d[\text{Ca}^{2+}]_{cyt}}{dt} &= 0.01[-3.768 \cdot 10^2\,(2F(I_{Ca\text{L}} + I_{Ca\text{T}}) + J_{PMCa}) + \\ &\quad + 2.1 \cdot 10^{-3}\Big(1,\,5 \cdot 10^{-4}\left([\text{Ca}^{2+}]_{ER} - [\text{Ca}^{2+}]_{cyt}\right) - J_{ERCa}\Big)\Big] \qquad (5.14)\\ \frac{d[\text{Ca}^{2+}]_{ER}}{dt} &= 2.1 \cdot 10^{-4}\left(-1,\,5 \cdot 10^{-4}\left([\text{Ca}^{2+}]_{ER} - [\text{Ca}^{2+}]_{cyt}\right) + J_{ERCa}\right).\end{aligned}$$

Here F is Faraday's constant, $[\text{Ca}^{2+}]_{ER}$ is the calcium in the ER, J_{PMCa}, J_{ERCa} are the Ca^{2+}-ATPase dependent efflux of Ca^{2+} ions via the plasma membrane and Ca^{2+} uptake by the ER-pumps, respectively. They are modelled by

$$\begin{aligned}J_{PMCa} &= M_{PMCa}\frac{[\text{Ca}^{2+}]_{cyt}^2}{[\text{Ca}^{2+}]_{cyt}^2 + K_{PMCa}^2} \qquad (5.15)\\ J_{ERCa} &= k_{ERCa}[\text{Ca}^{2+}]_{cyt}.\end{aligned}$$

Here M_{PMCa} is the maximal pump rate, K_{PMCa} is the half-maximal activation concentration for Ca^{2+}-ATPase, and k_{ERCa} is the ER pump parameter.

The model reproduces the spontaneous pacemaking and plateau bursting electrical activity on the discharge patterns of somatotrophs as well as the effects of changes in the extracelluar cation environment. Although receptors, ligand-receptor interactions and intracellular signalling pathways are not considered in the model, it does allow the implicit simulation of the effect of GHRH and SS on the induction of action potentials. These effects are achieved by a simultaneous increase/decrease in conductances for the Na^+ and inward-rectifier K^+ transmembrane channels.

At the time of writing, the absence of experimental data for the dynamics of ion channels *per se*, of the secondary sensory, PeVN and ArcN-ME neurons has meant that no mathematical models of their electrical activity have been developed. It is important to note that electrophysiological and pharmacological characterisation of ionic currents in general does not directly identify the ion channels involved. Single cell analysis is required to definitively link relative ion channel expression with their respective ionic currents and electrical activity patterns (Stojilkovic *et al* 2010). Based on their physiological requirements, i.e. the fast signal conduction and reliable sustained release of hormones, it can be speculated that they possess rapidly adapting and/or non-persistent stuttering spiking type activity, that could be described by

$$C_i\frac{\mathrm{d}V_i}{\mathrm{d}t} = -(\hat{I}_{\text{Ca}} + \hat{I}_{\text{Ca}-K} + \hat{I}_{\text{Na}} + \hat{I}_{\text{K}} + \hat{I}_{\text{Cl}}),\ (i = \text{SDH, PeVN, ArcN–ME}), \qquad (5.16)$$

where C_i is the membrane capacitance, $\hat{I}_{(...)}$ are the VGCa, BK_{Ca}, Na^+, K^+ and Cl^- ionic currents given by

$$\begin{aligned}
\hat{I}_{Ca} &= \frac{\hat{g}_{Ca} z}{1 + \zeta[Ca^{2+}]}\left(V_{nm} - \hat{V}_{Ca}\right) \\
\hat{I}_{Ca\text{-}K} &= \frac{\hat{g}_{Ca\text{-}K}\rho_\infty[Ca^{2+}]}{0.5 + [Ca^{2+}]}\left(V_{nm} - \hat{V}_{Ca\text{-}K}\right) \\
\hat{I}_{Na} &= \hat{g}_{Na}\hat{m}^3\hat{h}\left(V_{nm} - \hat{V}_{Na}\right)\tilde{g}_{Cl}\left(V_{nm} - \tilde{V}_K\right) \\
\hat{I}_{Cl} &= \hat{g}_{Cl}\left(V_{nm} - \hat{V}_{Cl}\right).
\end{aligned} \tag{5.17}$$

Here $\hat{V}_{(...)}$ are the reversal potentials for the respective currents, $\hat{g}_{(...)}$ are the maximal conductances, ζ is the parameter for calcium inhibition, z, ρ_∞, $\hat{m}$, $\hat{n}$, $\hat{h}$ are dynamic variables

$$\begin{aligned}
z &= (z_\infty - z)/\tau_z \\
\frac{d\hat{n}}{dt} &= (\hat{n}_\infty - \hat{n})/\tau_{\hat{n}} \\
\frac{d\hat{h}}{dt} &= \lambda_h(\hat{h}_\infty - \hat{h})/\tau_{\hat{h}} \\
\rho_\infty &= (1 + \exp 0.45(V_{nm} + 57))^{-1} \\
\hat{m} &= \hat{m}_\infty(V_{nm}).
\end{aligned} \tag{5.18}$$

Where

$$\begin{aligned}
\hat{y}_\infty &= \hat{\alpha}_{y\infty}\hat{\tau}_y \quad (\hat{y} = \hat{m}, \hat{h}, \hat{n}) \\
\hat{\tau}_y &= 1/(\hat{\alpha}_{y\infty} + \hat{\beta}_{y\infty})
\end{aligned} \tag{5.19}$$

$$\begin{aligned}
\hat{\alpha}_{\hat{m}\infty} &= \frac{0.12(V_{nm} + 27)}{1 - \exp(-(V_{nm} + 27)/8)}, \quad \hat{\beta}_{\hat{m}\infty} = 4.4\exp(V_{nm} + 47)/25 \\
\hat{\alpha}_{\hat{h}\infty} &= 0.09\exp(-(V_{nm} + 47)/27), \quad \hat{\beta}_{\hat{h}\infty} = (1 + \exp(-(V_{nm} + 33)/8))^{-1} \\
\hat{\alpha}_{\hat{n}\infty} &= \frac{0.012(V_{nm} + 12)}{1 - \exp(-(V_{nm} + 12)/8)}, \quad \hat{\beta}_{\hat{n}\infty} = 0.12\exp(-V_{nm} + 26)/87 \\
z_\infty &= (1 + \exp(-0.15(V_{nm} + 42)))^{-1}.
\end{aligned}$$

The dynamics of the free intracellular Ca^{2+} yields

$$\frac{d[Ca_i^{2+}]}{dt} = \frac{0.234 \times 10^{-4} z(V_{nm} - V_{Ca})}{1 + \zeta[Ca_i^{2+}]} - 0.003[Ca_i^{2+}]. \tag{5.20}$$

The values of constants and numerical parameters have been adjusted during numerical simulations to reproduce the desired electrophysiological responses.

5.2 Protein-tyrosine kinase receptor signalling

Until recently, signalling pathways have been viewed as relay routes that simply transmit information from the extracellular environment to intracellular effector sites. It has become increasingly clear that these involve many components subjected to multiple interlinked biochemical reactions and that their outcomes depend on the nature and characteristics of input stimuli and interaction kinetics among concomitant intermediates. Moreover, the crosstalk with other pathways that share similar components can influence the pattern of biological responses. A quantitative evaluation of these multifactorial and multiparametric processes is very difficult, and may even be impossible.

Different classes of mathematical models supported by specifically designed *in vivo* and *in vitro* experiments have been proposed to reproduce and study the intricacies of protein-tyrosine kinase receptor, e.g. GHR and IGF-1R, signalling. Two step sequential models include the structure of growth hormone binding surfaces (site 1 and 2) to its specific receptor. Site 1 has a high affinity to the extracellular domain of GHR, whilst site 2, responsible for homodimerisation of the receptor, has low affinity (Ilondo *et al* 1994, Wells 1996, Uchida *et al* 1999). The biological effect is directly proportional to the concentration of the fully activated GH-$(GHR)_2$ complex (here subscript 2 indicates dimerisation). Model parameters, the equilibrium dissociation constants for sites 1 and 2 (K_{d1}, K_{d2}) respectively, have been measured experimentally or adjusted during numerical runs to fit the recorded data. It has been shown that cell reaction to receptor activation exhibits a bell-shaped curve, i.e. the progressively increasing hormone concentration decreases the number of receptors available for dimerisation. The peak of the curve has been attributed to the initial binding of GH to a GHR while its broadness has been related to receptor dimerisation *per se* and intracellular signal transduction pathways. This modelling approach has also suggested a reduced formation of the nondimerized GH-GHR_1 complex which depends on the availability of free surface receptors and the affinity constant K_{d1}.

Haugh (2004), by incorporating trafficking processes, i.e. receptor internalisation and degradation, has advanced the ligand induced dimerisation model for the purpose of quantitative analysis of cell-based functional assay. The system of a differential equation of GHR internalisation (GHR_i) and three algebraic equations for nondimerized (C) and dimerised (D) complexes along with GHR has been derived based on the assumptions that: (i) the concentration of GH (L) remains constant throughout, (ii) the GH-$(GHR)_2$ dimerizarion is ordered, with site 1 of GH binding initially to form the GH-$(GHR)_1$ complex, (iii) active dimers are internalised at the same rate, and (iv) a steady-state is maintained through *de novo* GHR synthesis

$$\begin{aligned}\frac{d[GHR_i]}{dt} &= k_t([R] + [C]) - \left(k_{\text{rec}} + k_{\text{deg}}\right)[GHR]_i \\ [C] &= \frac{k_{f1}[L][R]}{k_{r1} + k_t + (k_{-x1} + k_e)K_X[R]}, \\ [D] &= \frac{k_{x2}}{k_{-x1} + k_{-x2} + k_e}[C][GHR] \\ [GHR] &= \frac{\left(k_{\text{deg}} + k_{\text{rec}}\right)(V_s - 2k_e[D])}{k_t k_{\text{deg}}} - [C].\end{aligned} \tag{5.21}$$

The initial conditions are given by $C_0 = D_0 = 0, [GHR_0] = V_s(k_{\text{deg}} + k_{\text{rec}})/k_t k_{\text{deg}}$, and $k_{f1}, k_{r1},$, $k_{-x1}, k_{\pm x2}$, $k_{\text{deg}}, k_{\text{rec}}, k_e, k_t$, Vs, are rate constants of forward and reverse reactions of the GH-GHR1 and GH-GHR2 complexes formation, degradation, recycling and internalisation, and GHR synthesis, respectively, K_X is the cross-linking constant.

The dimerised GH-GHR_2 complex brings into close proximity the JAK2 kinases and causes their rapid transphophorylation with activation of cytosolic signalling pathways following. The strength of proliferation signal (S) is modelled by

$$S = \frac{2D/R_0}{k_E + 2D/R_0} \tag{5.22}$$

which is an analogue of the Hill equation.

The model was tested to match the dose–response curves for an adjusted set of parameter values for site 1 and 2 receptor binding affinity. The results of investigations indicate the possibility of antagonists to the GHR by introducing mutations in GH that block binding at site 2 (reducing the value of K_X) and even increasing their pharmacological potency by simultaneously enhancing affinity at site 1.

Assuming that IGF-1 induced cell proliferation is independent of insulin receptor substrate activity, a mass-action kinetics mathematical model of IGF-1 interaction with cognare receptors and specific binding proteins was developed to study the dose response effect of immortalised ovarian cancer cells to treatment by IGF-1R antibodies and freely available IGFBPs (Tian and Kreeger 2014). The results of modelling, confirmed by experiments designed and conducted *in vitro*, have demonstrated a saturation response proportional to the maximal number of occupied and activated IGF-1Rs thus suggesting that antibodies against IGF-1Rs might be more therapeutically effective globally.

A detailed deterministic approach to investigate the cell line-specific recruitment of Shc2 homology proteins and phosphotyrosine-binding domains to multiple cytoplasmic autophosphorylaton sites of the IGF-1R has been proposed by Erickson *et al* (2019). In the construction of this model it has been assumed that IGF-1Rs exist in pre-formed dimeric forms, each containing two functionally equivalent pockets and each with two dissimilar binding sites (S_1 and S_2). The latter interact reversibly in a sequential manner with IGF-1 to form the cross-linked

S_1: IGF-1:S_2 complex that stabilises active conformation. IGF-receptors possess six autophophorylation sites and allow the recruitment of multiple proteins simultaneously. Their kinetics is described by a cyclic reaction scheme with eight defined reactions that satisfy the principle of detailed balance (Alberty 2004, Yang *et al* 2006). The results of model investigations have offered new mechanisms of competitive protein recruitment in receptor-mediated signalling, i.e. the rank in which a substrate is recruited does not depend entirely on its binding affinity but rather on its abundance profile. This implies that pathway selectivity could be the consequence of expression levels of the proteins involved and their avidity for the activated receptor *per se*.

A synergistic *in silico* and *in vitro* approach to quantify short term signalling in rat hepatocytes by the epidermal growth factor protein-tyrosine kinase receptor (EGFR) through its interaction with Shc, PLCγ, Grb2 (growth factor receptor binding 2) and SOS (Son of Sevenless) proteins has been put forward by Kholodenko *et al* (1999) in the analysis of the kinetics of the EGFR pathway. The processes modelled include: (i) the sequential formation of the monomeric and dimeric EGF–EGFR reactive complexes, (ii) the competitive binding of cytoplasmic Shc, PLCγ, Grb2 and SOS proteins to the phosphorylated EGF–EGFR complex, and (iii) the complete cycle of chemical reactions of Shc, PLCγ, Grb2 and SOS pathways leading to, but not considered here, the Ras-protein cascade. The numerical simulations identify the kinetic properties of reactions responsible for the transient behaviour of the system. Thus, it shows that the rate of EGFR tyrosine phosphorylation is strictly controlled by the reaction of binding of Shc and/or PLCγ to the receptor and the activity of phosphatase(s), both regulated by the amount of phosphorylated EGFR.

Over the last two decades a number of mathematical models have been proposed which differ in complexity and detail to investigate the various adaptive topologies of RasMAPK-signalling in the context of diverse Ras isoforms, cellular localisation and disparate backgrounds (Markevich *et al* 2004, Brightman and Fell 2000, Schoeberl *et al* 2002, Kholodenko *et al* 1999, Orton *et al* 2005, Stites *et al* 2007, Orton *et al* 2009, Stites *et al* 2015). These models have mostly focused on RAS activation within a single receptor tyrosine kinase pathway, neglecting the impacts of positive and negative interactions among multiple RAS effector networks, e.g. MAPK/ERK1/2. Such feedback loops can produce unexpected dynamic behaviours such as the occurrence of relaxation oscillations, the formation of travelling phosphorylation waves, bistable modes of action, etc (Markevich *et al* 2006, Kiyatkin *et al* 2006, Qiao *et al* 2007, Ritt *et al* 2010). Schoeberl *et al* (2002) developed an integrated quantitative, dynamic, and topologically detailed model of epidermal growth factor (EGF) receptor intracellular signal networks. The results of numerical simulations reveal that the rate of receptor activation plays a critical role in signal transduction efficacy and stability of the system with regard to input data, i.e. ligand concentrations. Das *et al* (2004) in their mathematical model of SOS-mediated activation of T cell membrane-bound Ras proteins assumed that the SOS protein binds to RasGDP and RasGTP, respectively, to form SOS–RasGDP and

SOS–RasGTP complexes. Cyclic conversion between them is catalised by GTPase kinase. The system of four ordinary differential equations was derived and solved numerically. It should be noted that none of the concentration values and only a few of the model parameters that were used are known for T lymphocytes. These have been estimated from various sources in literature and/or obtained from stochastic simulations and computational parameter sensitivity tests. With this 'uncertain' set of input data, the authors have demonstrated the possible existence of stable Ras activity in relation to high or low concentrations of SOS. However, for intermediate levels of SOS, a bistability of the dose–Ras response curve is predicted which has been confirmed experimentally (Lun *et al* 2017, Birtwistle *et al* 2012, Dobrzynski *et al* 2014, Dobrzynski *et al* 2012, Shin *et al* 2008, Kochańczyk *et al* 2017). Interestingly, this can be abrogated if the formation of SOS–RasGDP is excluded from consideration. Based on experimental data of protein copy numbers and their binding affinities of eleven cell lines, Stites *et al* (2015) developed a rule-based model of epidermal growth factor receptor signalling to investigate potential cell-line specific differences in the rank order of recruited EGFR binding partners.

Despite great enthusiasm and promises shared amongst computational biologists, current *in silico* approaches of Ras/MAPK signalling bear a low level of reliability and feasibility. The meticulous reproduction of the cascades of biochemical reactions drags in mathematical models unnecessary complexity without the desired accuracy and Gnostic properties to the model. No attempts have been made so far to integrate the receptor-signalling pathway-effector response events in a unified model to study the dynamics of their convoluted behaviour.

5.3 Gene expression

Gene expression encompasses processes by which information encoded in DNA is copied (transcribed) to mRNA and in turn decoded (translated) in the ribosomes during protein synthesis. These comprise multiple interacting factors, co-factors and proteins linked via biochemical reactions to form a dynamic gene regulatory network (GRN). Being highly stochastic and subjected to external/internal pertubations, the GRN is difficult to assess intuitively alone.

Two broad classes of mathematical models, stochastic and analytical, have been developed to study the GRN. The former employ general principles of neural and Bayesian networks and have proved invaluable in analysing large-scale gene data sets, estimating protein production and their variance on transcription and translation rates, and determining and evaluating the characteristic parameters of the system (Robert 2019, Wan *et al* 2019, Rothenberg 2019, Cambón and Sánchez 2019). However, these models cannot explain the fine details of binding of RNA polymerase to DNA, cooperative and inhibitory interactions between transcription factors, the mRNA translation rate, protein degradation, etc. These are effectively captured by analytical (thermodynamic, Boolean and deterministic) models. In the thermodynamic approach, also known as fraction occupancy, each possible state ($j = 1, \ldots, n$) of the enhancer is assigned a statistical weight, w_j, as a function of the concentration and binding affinity of available transcription factors. Given the probability of each state, i.e. transcribed or silent, $P_j = w_j / \sum_{j=1}^{n} w_j$, gene expression output is calculated as either proportionate to the transcribed probability or as the weighted sum of transcription factors (Bintu *et al* 2005, He *et al* 2010). Thus, this approach drastically simplifies the complexity of DNA-transcription factor interactions to consider only the binding, recruitment and facilitation of transcription and its co-factors. Despite their constructive limitations, these models have helped: detect and predict new transcription factor targets effectively (Granek and Clarke 2005); analyse cooperative and competitive interactions between various transcription factors and binding sites (Gertz and Kohen 2009); explore and explain the variance in gene expression depending on promoter structures (Segal *et al* 2009, Gertz *et al* 2009, Gertz and Kohen 2009); investigate in detail the function of a single regulatory region within the GRN (Reinitz *et al* 1995, Papatsenko *et al* 2015).

Biological problems require models that reconstitute multicomponent, spatially and temporally evolving dynamic systems using the general laws and principles of physics and measurable parameters. From the pioneering publications of deterministic mathematical models for the *lac* (lactose) and *trp* (tryptophan) operon dynamics by Goodwin (1965) and Griffith (1968a, 1968b), there has followed an explosion of *in silico* research into the broad spectrum of gene expression problems such as the dynamics of *Bcd* (Bicoid), a simple diffusible morphogen, (Jaeger *et al* 2004a), complex cell-to-cell GRNs (Mackey *et al* 2004, Hirata *et al* 2002, Smolen *et al* 2000, Vilar *et al* 2003, Wang *et al* 2012, Saadatpour and Albert 2012, Hancock *et al* 2015), promoter decoding and translation (Choubey 2018, Hansen and O'Shea 2013, Jaeger *et al* 2004b; Keränen *et al* 2006, Perez-Ortin *et al* 2007, Sen and Liu 1990, Skinnet *et al* 2016), gap gene networks (Reinitz and Sharp 1995), stability analysis of the *lac* and *trp*-operon systems (Bliss *et al* 1982, Santillán and Mackey 2004, Santillán 2008, Yildirim and Mackey 2003, Angelova and Ben-Halim 2011), among others. As the number of such publications is constantly growing, the interested reader is advised to consult specialised literature.

Most of these models are based on the law of mass action (Michaelis–Menten) and the diffusion reaction type equations with reacting components and kinetic constants describing biochemical processes as the main variables and input data. Although showing their robustness in studying dynamic aspects of gene expression and even predicting effects before being confirmed experimentally, the continuous abstraction, the fundamental implicit assumption in the differential approach, is not always warranted when dealing with molecular stochastic systems. Moreover, lack of reliable *in vivo* and *in vitro* information affects significantly the accuracy, and consequently, the practicality of simulation results. Therefore, combined stochastic-deterministic, thermodynamic-deterministic and Boolean models are those being successfully developed and implemented in studies of gene expression (Jaeger *et al* 2004a, 2004b, Angelova and Ben-Halim 2011, Huang *et al* 2018).

At the time of writing, no research publications were available on gene expression related to collagenesis or myogenesis of the bladder wall.

5.4 Hormonal interactions

Information on the anatomy, biochemistry and physiology of GH, SS, GHRH and IGF-1 (see above), addresses mainly qualitative aspects of hormone production, secretion and action whereas time-dependent characteristics of the processes and mechanisms of their interactions are omitted. Although existing empirical quantitative data on $k_{\pm}$ and $t_{1/2}$ has allowed statistical correlations to be established between pulsatile bursts of GH as a function of an ultradian rhythm, such data cannot provide an insight into the convoluted dynamics of interplay among the reacting substrates (Veldhius *et al* 1987).

By contrast, the deterministic approach offers an opportunity for a rigorous analysis and interpretation, in a physiologically meaningful context, of the chemical reactions involved and in providing a quantitative assessment of their dynamics. A series of such models have been developed to reproduce and study pituitary-(intra)hypothalamic relations in the GH axis. The first elaborate physiologically accurate kinetic description of the GH control system to regulate expressions for synthesis, along with the coordinated release and elimination of GH, SS, GHRH and IGF-1, was developed by Chen *et al* (1995). It comprises of the PeVN, ArcN-ME complex, adenohypophysis and the liver arranged in a network interlinked through stimulatory and inhibitory feedback loops. The working assumptions employed in the mathematical formulation are: (i) the chemical reactions of first-order kinetics, (ii) the excitatory (superscript +) and inhibitory (superscript −) signalling, described phenomenologically by the modifying functions $F_{iXj}^{+/-}$(adapted Hill function)

$$F_{iXj}^{+/-}(D) = \frac{u - v}{1 + (D/D_t)^p} + v, \quad u = F_{iXj}^{+/-}(0), \quad v = F_{iXj}^{+/-}(\infty), \tag{5.23}$$

where $i, j = \overline{1, 4}$ corresponds to GHRH, SS, GH and IGF-1, respectively, X refers either to the primary process of release, $X = R$, or $X = S$ - synthesis. Thus, $F_{1(X=R)3}^{+/-}$ indicates the stimulatory effect of hypothalamically released hypophysial portal GHRH ($i = 1$) on pituitary GH ($j = 3$) release. Cumulative biological effects of primary processes are implemented in the model in a simple multiplicative manner whilst dual regulation, i.e. conjoint stimulation and inhibition, is treated as analogous to the classic enzyme kinetic approach to model the competitive and non-competitive ligand-substrate action. The delay variable D (D_t- the threshold value) is defined by

$$D(t) = \frac{1}{\tau}\int_{-\infty}^{t-d} H(z)\exp\left[-\frac{(t - d) - z}{\tau}\right]dz. \tag{5.24}$$

Here $H(z)$ is the concentration of hormone at time z, d is the time delay, and τ is the duration period of the stimulus. Based on the application of the law of mass action to the chain of GHRH, SS and GH neuroendocrine chemical reactions, the coupled system of ordinary differential equations is derived thus

$$\frac{dS_i}{dt} = k_{Si}\Phi_i\left(F_{iXj}^{+/-}(D_{iXj})\right)(S_i^m - S_i) - RS_i - RS_{ih}$$

$$\frac{dH_i}{dt} = \frac{RS_i}{V_1} - k_{Ei}H_{ih} \tag{5.25}$$

$$\frac{dH_{jh}}{dt} = RS_{jh} - k_{Ejh}H_{jh}, \ (i = 1, 2, 3; j = 1, 2).$$

Here S_i^m, S_i, S_{jh} are the maxima and current concentrations of a stored hormone, and H_i, H_{ih}, H_{jh} are the amount of hormone released from the hypophysis and intrahypo-thalamus, respectively, k_{Si}, k_{Ei}, k_{Eih} are the chemical rate constants, and $RS_{i(jh)}$ are the regulatory functions, $RS_{i(jh)}=\Psi_{i(jh)}(F_{iXj}^{+/-}(D_{iXj}))$. Their structure depends entirely on the type and sequence of reactions considered. The model contains 72 parameters, the majority of which are unmeasurable. Therefore, these have to be either guessed or adjusted during numerical runs to bring outputs to the best fit with experimental data. As such the model reproduces the dynamic behaviour of GH, i.e. rhythmicity, changes in the concentration profile and bursts of release, in response to exogenous infusion and withdrawal of regulatory hormones. These result from defined *a priori* infusion and withdrawal functions that take into account both the rates, duration and time of onset/offset of infusion/withdrawal. The model is capable of reproducing the periodicity of GH secretion under normal physiological conditions and after imposed pertubations. This includes the partial or full withdrawal of GHRH, the infusion of exogenous GH with/without subsequent withdrawal of SS, the infusion of SS and GH and the concomitant withdrawal of GHRH, and injections of GHRH during maxima/ minima concentrations of GH with/without prior SS withdrawal.

A conceptually similar methodology with some modifications in its approach has been employed by other researchers. Thus, Wagner *et al* (1998) have simulated specifically the 3.3-hr two-peak ultradian rhythm of GH secretion observed in male rats. It has been hypothesised that the dual feedback-control mechanism of GH release by somatotrophs involves: (i) the direct activation by GH of GH-R on SS containing PeVN neurons, and (ii) a delayed SS spill into the portal circulation and the brain. The following assumptions have been made with regard to GH-R kinetics: (i) upon binding to GH, these quickly achieve and remain in a steady-equilibrium state throughout, (ii) the release of SS from PeV neurons is proportional to the concentration of 'internalised' GH-Rs; (the term 'internalisation' is a misuse, although authors, rightfully, connote the amount of biologically active substrate), and (iii) the cooperative hormone binding to the receptor is described by the Hill function. As in the Chen model, the combined regulatory effects of ligands are represented by an appropriate combination of Hill functions. The excitation to the system is provided by GHRH which is released at regular periods of ~1 hr independently of GH. The mathematical formulation of the model is given by two differential equations describing the time course of GH and SS and contains 14 parameters. These are calibrated in such a way that the simulated concentrations and timing are comparable to experimentally measured values and recordings.

A series of publications has been dedicated to the mathematical modelling and analysis of species and gender dependent secretion of GH (Farhy *et al* 2001, 2002, Farhy and Veldhuis 2003, 2004). Operating within the framework of Chen's formalism, Farhy *et al* have constructed a network of five principal dose–responsive forward/feedback regulatory functions that describe: (i) the time-delayed reciprocal interactions between GH and GHRH generating high frequency pulsatile GH release, (ii) the time-lagged bidirectional GH-SS interplay with inherent low frequency GH bursts, and (iii) intrahypothalamo-hypothalamo-pituitary-somatotroph interactions that drive GH secretion. Specifically, the aggregate hypothalamo-pituitary effects are subsumed in the model by a stochastic term which relates random variability in the sensitivity of GHRH containing ArcN neurons to the total amount of SS present in the system. The mathematical formulation employs 15 parameters most of which have been estimated from *in vivo* and *in vitro* experiments on rats. Only a few variables related to the Hill functions have been adjusted during simulations. Even with this constructive 'uncertainty', the results of numerical experiments recapitulate the complex gender dependent patterns of coordinated release of hormones. Thus, the GH-induced SS secretion and the damping of slower frequency oscillatory activity produces high concentration levels of GH similar to those observed in male rats. By contrast, the increase in SS outflow and the subsequent inhibition of high frequency oscillations converts the mode of GH bursts to that generated by rats of opposite gender.

In the deterministic quantitative pituitary model of GH secretion developed by Brown *et al* (2004) it is assumed that the release of GH is GHRH-dependent and/or independent (constitutive secretion). GHRH-induced release occurs through the activation of effector elements, analogues of GHRH-Rs, located at somatotrophs. Although the role of GHRH-Rs in the GH axis is strongly emphasised, the receptors and linked intracellular signalling pathways are both treated phenomenologically as a single entity. The down-regulation of GH and GHRH secretion is through SS secreted by PeV neurons. The dynamics of hormones is described by functions constructed to guarantee the best fit with the author's *in vitro* experiments.

MacGregor and Leng (2005) elaborated the model to serve their experiments on pulsatile secretion of GH. In their study they hypothesise that the GHRH pulse generator is a result of interactions between PeVN and ArcN neurons which are also under the influence of free circulating GH. The latter provides a delayed, long-lasting internal feedback in the system to control somatostatin release. Although GH does not directly trigger SS secretion, it does augment its stores in proportion to the concentration of GH present at an earlier time. Thus, when the stores are empty there is no SS release, the secretion occurring only when the stores are full. The input in the form of either regular or random (Poissonian) perturbation acts as a signal for the release. All actions of peptides as in previous models are given by the Hill equations or their convolutions. The complete mathematical formulation is given by

$$\begin{aligned}
\frac{dr}{dt} &= r_{act}h_rI_r - k_6r + I_{r2},\\
\frac{ds}{dt} &= I_sg - k_7s + I_{s2},\\
\frac{ds_e}{dt} &= I_{se} - k_7s_e,\\
\frac{df}{dt} &= -k_1(r + c)f + [k_2 + k_3\Phi_r(s + s_e)](1 - f),\\
\frac{dh_h}{dt} &= j_1(I_h - k_8h),\\
\frac{dg}{dt} &= g_1h_{act}[(g_{\max} - g)/g_{\max}] - g_2I_2g,\\
\frac{dh}{dt} &= j_1[(k_4 + k_5[1 - \Phi_b(s + s_e)])(r + c)f - k_8h],\\
r_{act} &= 1 - s^{n1}/(s^{n1} + \theta_1^{n1}),\\
h_{act} &= h_{\Delta}^{n2}/\left(h_{\Delta 1}^{n2} + (j_1\theta_2)^{n2}\right),\\
h_r &= 1 - h_{\Delta 1}^{n3}/\left(h_{\Delta 1}^{n3} + (j_1\theta_3)^{n3}\right),\\
\Phi_b(s) &= 1/\left[1 + \exp(-\left[\log_{10}(s) - s_b\right]/\delta_b\right],\\
\Phi_r(s) &= 1/\left[1 + \exp(-\left[\log_{10}(s) - s_r\right]/\delta_r\right],\\
h_{\Delta} &= (h + h_h)\Delta.
\end{aligned} \tag{5.26}$$

Here

$$r = [GHRH],\ s_{(e)} = [SS_{(e)}],\ h_{(h)} = [GH_{(h)}],\ g = [SS_{\text{pool}}],\ f = [GHRH - R]$$

where the subscript (e) is referred to external and (h)—human. Other parameters are explained thus: $g_{(1,\,2,\,\max)}$ are scaling factors related to SS release; j_1 is the scaling factor fitting the GH concentration to *in vivo* levels; $n_{1,\,2}$ are gradient constants of inhibition of GHRH and stimulation of GH secretion by SS, respectively, with n_3 as the gradient of down-regulation of GHRH release by GH; k_{1-8} are rates of chemical reactions; $\theta_{1,\,2}$ are threshold values of inhibition of ArcN neurons and stimulation of somatotrophs, respectively; θ_3 is the threshold level of inhibition of ArcN neurons by GH; $I_{(r,s)}$ are the release rates of GHRH and SS; $I_{(r2,s2,h)}$ are concentrations of injected GHRH, SS and GH; $\Delta_{(1)}$ are time delay constants of the stimulation of SS release and the inhibition of GH secretion by GH; c is the level of constitutive activation of GHRH-Rs; $s_{b,r}$ are threshold values of activation of GHRH-R recycling and inhibition of GH secretion, respectively; $\delta_{b,r}$ are gradients of SS effects.

The models, as described above, are derived to reproduce particular experiments conducted by the authors themselves and, as such, are fragmentary and incomplete, i.e. serving only to demonstrate the existence of hormonal interplay within the hypothalamo-pituitary axis without unravelling the details of its dynamic interactions with the effector organ(s). These models: (i) lack Gnostic capabilities and intrinsic flexibility, the essential inherent property of each model, (ii) are not general enough to offer a coherent description of essential biochemical reactions within the unified framework, (iii) are 'parameter' biased according to the experimental conditions they replicate, and therefore, (iv) render the intractable failing to offer an explanation for the complex processes operating among GH, GHRH, SS and IGF-1 and their downstream intracellular pathways in terms of spatiotemporal biological representations.

References

Alberty R A 2004 Principle of detailed balance in kinetics *J. Chem. Educ.* **81** 1206–9

Andrietti F and Bernardini G 1984 Segmented and 'equivalent' representation of the cable equation *Biophys. J.* **46** 615–23

Angelova M and Ben-Halim A 2011 Dynamic model of gene regulation for the lac operon *J. Phys. Conf. Ser.* **286** 012007

Babbs C F and Shi R 2013 Subtle paranodal injury slows impulse conduction in a mathematical model of myelinated axons *PLoS One* **8** e67767

Basser P J 1983 Cable equation for a myelinated axon derived from its microstructure *Med. Biol. Eng. Comput.* **31** S87–92

Binczak S, Eilbeck J C and Scott A C 2000 Ephaptic coupling of myelinated nerve fibers *Phys. D: Nonlinear Phenom.* **148** 159–74

Bintu L, Buchler N E, Garcia H G, Gerland U, Hwa T, Kondev J, Kuhlman T and Philips R 2005 Transcriptional regulation by the numbers: models *Curr. Opin. Genet. Dev.* **15** 116–24

Birtwistle M R, Rauch J, Kiyatkin A, Aksamitiene E, Dobrzy M, Hoek J B, Kolch W, Ogunnaike B A and Kholodenko B N 2012 Emergence of bimodal cell population responses from the interplay between analog single-cell signaling and protein expression noise *BMC Syst. Biol.* **6** 109

Bliss R D, Painter P R and Marr A G 1982 Role of feedback inhibition in stabilizing the classical operon *J. Theor. Biol.* **97** 177–93

Brightman F A and Fell D A 2000 Differential feedback regulation of the MAPK cascade underlies the quantitative differences in EGF and NGF signalling in PC12 cells *FEBS Lett.* **482** 169–74

Brill M H, Waxman S G, Moore J W and Joyner R W 1977 Conduction velocity and spike configuration in myelinated fibres: computed dependence on internode distance *J Neurol Neurosurg. Physch.* **40** 769–74

Brown D, Stephens E A, Smith R G, Li G and Leng G 2004 Estimation of parameters for a mathematical model of growth hormone secretion *J. Neuroendocrinol.* **16** 936–46

Burridge R and Keller J 1981 Poroelasticity equations derived from microstructure *J. Acoust. Soc. Am.* **70** 1140–6

Cambón M and Sánchez Ó 2019 Analysis of the transcriptional logic governing differential spatial expression in Hh target genes *PLoS One* **14** e0209349

Chen L, Veldhuis J D, Johnson M L and Straume M 1995 System-level analysis of physiological regulatory interactions controlling complex secretory dynamics of the growth hormone axis: a dynamical network model *Methods in Neuroscience* (New York: Academic) pp 270–310

Choubey S 2018 Nascent RNA kinetics: transient and steady state behaviour of models of transcription *Phys. Rev.* E **97** 022402

Das J, Ho M, Zikherman J, Govern C, Yang M, Weiss A, Chakraborty A K and Roose J P 2004 Digital signalling and hysteresis characterise Ras activation in lymphoid cells *Cell* **136** 337–51

De Young G W and Keizer J 1992 A single pool IP3-receptor based model for agonist stimulated Ca^{2+} oscillations *Proc. Natl Acad. Sci. USA* **89** 9895–9

Dobrzynski M, Fey D, Nguyen L K and Kholodenko B N 2012 Bimodal protein distributions in heterogeneous oscillating systems *CMSB 2012 (Lecture Notes in Computer Science* vol 7605) (Berlin: Springer) pp 17–28

Dobrzynski M, Nguyen L K, Birtwistle M R, von Kriegsheim A, Fernandez A B, Cheong A, Kolch W and Kholodenko B N 2014 Nonlinear signalling networks and cell- to cell variability transform external signals into broadly distributed or bimodal responses *J. Roy. Soc. Interf.* **11** 20140383

Dutta D J *et al* 2018 Regulation of myelin structure and conduction velocity by perinodal astrocytes *Proc. Natl Acad. Sci. USA* **115** 11823–37

Erickson K E *et al* 2019 Modelling cell line-specific recruitment of signalling proteins to the insulin-like growth factor 1 receptor *PLOS Comp. Biol.* **15** e1006706

Farhy L S, Straume M, Johnson M L, Kovatchev B and Veldhius J D 2001 A construct of interactive feedback control of the GH axis in the male *Am. J. Physiol.* **281** R38–51

Farhy L S, Straume M, Johnson M L, Kovatchev B and Veldhuis J D 2002 Unequal autonegative feedback by GH models the sexual dismorphism in GH secretory dynamics *Am. J. Physiol. (Reg. Integr. Comp. Physiol.)* **282** R753–64

Farhy L S and Veldhuis J D 2003 Joint pituitary-hypothalamic and intrahypothalamic autofeedback construct of pulsatile growth hormone secretion *Am. J. Physiol. (Reg. Integr. Comp. Physiol.)* **285** R1240–9

Farhy L S and Veldhuis J D 2004 Putative GH pulse renewal: periventricular somatostatinergic control of an arcuate-nuclear somatostatin and GH-releasing hormone oscillator *Am. J. Physiol. (Reg. Integr. Comp. Physiol.)* **286** R1030–42

Fitzhugh R 1962 Computation of impulse initiation and saltatory conduction in a myelinated nerve fiber *Biophys. J.* **2** 11–21

Freeman S A, Desmazieres A, Fricker D, Lubetzki C and Sol-Foulon N 2016 Mechanisms of sodium channel clustering and its influence on axonal impluse conduction *Cell. Mol. Life Sci.* **73** 723–35

Gertz J and Kohen B A 2009 Environment-specific combinatorial cis-regulation in synthetic and genomic promoters *Mol. Syst. Biol.* **5** 244

Gertz J, Siggia E D and Kohen B A 2009 Analysis of combinatorial cis-regulation in synthetic promoters *Mol. Syst. Biol.* **5** 244

Goodwin B 1965 Oscillatory behavior in enzymatic control process *Adv. Enz. Regul.* **3** 425–38

Granek J A and Clarke N D 2005 Explicit equilibrium modeling of transcription factor binding and gene regulation *Nature* **457** 215–8

Griffith J S 1968a Mathematics of cellular control processes. I. Negative feedback to one gene *J. Theor. Biol.* **20** 202–8

Griffith J S 1968b Mathematics of cellular control processes. II *Positive feedback to one gene J. Theor. Biol.* **20** 209–16

Hancock E J, Stan G-B, Arpino J A J and Papachristodoulou A 2015 Simplified mechanistic models of gene regulation for analysis and design *JRS Interface* **12** 20150312

Hansen A S and O'Shea E K 2013 Promoter decoding of transcription factor dynamics involves a trade-off between noise and control gene expression *Mol. Syst. Biol.* **9** 704

Haugh J M 2004 Mathematical model of human growth hormone (hGH)-stimulated cell proliferation explains the efficacy of hGH variants as receptor agonists or antagonists *Biotechnol. Prog.* **20** 1337–44

He X, Samee M A, Blatti C and Sinha S 2010 Thermodynamic-based models of transcriptional regulation by enhancers: the roles of synergistic activation, cooperative binding and short-range repression *PLoS Comput. Biol.* **6** e10000935

Hirata H, Yoshiura S, Ohtsuka T, Bessho Y, Harada T, Yoshikawa K and Kageyama R 2002 Oscillatory expression of the bHLH Hes1 regulated by a negative feedback loop *Science* **298** 840–3

Hodgkin A L and Huxley A F 1952 A quantitative description of membrane current and its application to conduction and excitation in nerve *J. Physiol.* **117** 500–44

Huang B, Jia D, Feng J, Levine H, Onuchic J N and Lu M 2018 RACIPE: a computational tool for modelling gene regulatory circuits using randomization *BMC Sys. Biol.* **12** 74

Ilondo M M, Damholt A B, Cunningham B C, wells J A, De Meyets P and Shymko R M 1994 Receptor dimerization determines the effects of growth hormone in primary rat adipocytes and cultured human IM-9 lymphocytes *Endocrine* **134** 2397–403

Jack J J B, Noble D and Tsien R W 1975 *Electrical Current Flows in Excitable Cells* (Oxford: Clarendon)

Jaeger J *et al* 2004b Dynamical analysis of regulatory interactions in the gap gene system of *Drosophila melanogaster Genetics* **167** 1721–37

Jaeger J *et al* 2004a Dynamic control of positional information in the early Drosophila embryo *Nature* **430** 368–71

Keränen S V, Fowlkes C C, Luengo Hendriks C L, Sudar D, Knowles D W, Malik J and Biggin M D 2006 Three-dimensional morphology and gene expression in the *Drosophila* blastoderm at cellular resolution II: dynamics *Genome. Biol.* **7** R124

Kholodenko B N, Demin O V, Moehren G and Hoek J B 1999 Quantification of short term signaling by the epidermal growth factor receptor *J. Biol. Chem.* **274** 30169–81

Kiyatkin A, Aksamitiene E, Markevich N I, Borisov N M, Hoek J B and Kholodenko B N 2006 Scaffolding protein Grb2-associated binder 1 sustains epidermal growth factor-induced mitogenic and survival signaling by multiple positive feedback loops *J. Biol. Chem.* **281** 19925–38

Kochańczyk M, Kocieniewski P, Kozłowska E, Jaruszewicz-Błońska J, Sparta B, Pargett M, Albeck J G, Hlavacek W S and Lipniacki T 2017 Relaxation oscillations and hierarchy of feedbacks in MAPK signalling *Sci. Rep.* **7** 38244

Lun X, Zanotelli V R T, Wade J D, Schapiro D, Tognetti M, Dobberstein N and Bodenmiller B 2017 Influence of node abundance on signaling network state and dynamics analyzed by mass cytometry *Nat. Biotechnol.* **35** 164–72

MacGregor D J and Leng G 2005 Modelling the hypothalamic control of growth hormone secretion *J. Neuroendocrin.* **17** 788–803

Mackey M C, Santillán M and Yildirim N 2004 Modeling operon dynamics: the tryptophan and lactose operons as paradigms *Cell Res. Biol.* **327** 211–24

Markevich N I, Moehren G, Demin O V, Kiyatkin A, Hoek J B and Kholodenko B N 2004 Signal processing at the Ras circuit: what shapes Ras activation patterns? *Syst. Biol.* **1** 104–13

Markevich N I, Tsyganov M A, Hoek J B and Kholodenko B N 2006 Long-range signaling by phosphoprotein waves arising from bistability in protein kinase cascades *Mol. Syst. Biol.* **2** 61

Masoli S, Rizza M F, Sgritta M, van Geit W, Schürmann F and D'Angelo E 2017 Single neuron optimization as a basis for accurate biophysical modeling: the case of cerebellar granule cells *Front Cell Neurosci.* **11** 71

Nfor N O and Mokoli M T 2016 Dynamics of nerve pulse propagation in a weakly dissiative myelinated axon *J. Mod. Phys.* **7** 1166–80

Orton R J, Adriaens M E, Gormand A, Sturm O E, Kolch W and Gilbert D R 2009 Computational modelling of cancerous mutations in the EGFR/ERK signalling pathway *BMC Syst. Biol.* **3** 100

Orton R J, Sturm O E, Vyshemirsky V, Calder M, Gilbert D R and Kolch W 2005 Computational modelling of the receptor-tyrosine-kinase-activated MAPK pathway *Biochem. J* **392** 249–61

Papatsenko D, Darr H, Kulakovskiy I V, Waghray A, Makeev V J, MacArthur B D and Lemischka I R 2015 Single-cell analyses of ESCs reveal alternative pluripotent cell states and molecular mechanisms that control self-renewal *Stem Cell Rep.* **5** 207–20

Perez-Ortin J, Alepuz P M and Moreno J 2007 Genomics and gene transcription kinetics in yeast *Trend Gen.* **23** 250–7

Qiao L, Nachbar R B, Kevrekidis I G and Shvarstman S Y 2007 Bistability and oscillations in the Huang-Ferrell model of MAPK signaling *PLoS Comput. Biol.* **3** e184

Reinitz J, Mjolsness E and Sharp D H 1995 Model for cooperative control of positional information in Drosophila by bicoid and maternal hunchback *J. Exp. Zool.* **271** 47–56

Reinitz J and Sharp D H 1995 Mechanism of *eve* stripe formation *Mech. Dev.* **49** 133–58

Ritt D A, Monson D M, Specht S I and Morrison D K 2010 Impact of feedback phosphorylation and raf heterodimerization on normal and mutant B-Raf signalling *Mol. Cell. Biol.* **30** 806–19

Robert P 2019 Mathematical modelling of gene expression arXiv:1905.02578v3 [q-bio.MN]

Rothenberg E V 2019 Causal gene regulatory network modeling and genomics: second-generation challenges *J. Comput. Biol.* **26** 703–18

Saadatpour A and Albert R 2012 Discrete dynamic modeling of signal transduction networks *Methods Mol. Biol.* **880** 255–72

Santillán M 2008 Bistable behaviour in a model of the lac operon in *Escherichia coli* with variable growth rate *Biophys. J.* **94** 2065–81

Santillán M and Mackey M C 2004 Influence of catabolite repression and inducer exclusion on the bistable behavior of the lac operon *Biophys. J.* **86** 1282–92

Schierwagen A and Ohme M 2008 A model for the propagation of action potentials in nonuniform axons *AIP Conf. Proc.* **1028** 98

Schoeberl B, Eichler-Jonsson C, Gilles E D and Müller G 2002 Computational modeling of the dynamics of the MAP kinase cascade activated by surface and internalized EGF receptors *Nat. Biotechnol.* **20** 370–5

Segal E, Raveh-Sadka T, Schroeder M, Unnerstall U and Gaul U 2009 Predicting expression patterns from regulatory sequence in Drosophila segmentation *Nature* **451** 535–40

Sen A K and Liu W 1990 Dynamic analysis of genetic control and regulation of amino acid synthesis: the *tryptophan* operon in *Escherichia coli Biotech. Bioeng.* **35** 185–94

Shin S, Rath O, Choo S, Fee F, McFerran B, Kolch W and Cho K 2008 Positive- and negative-feedback regulations coordinate the dynamic behavior of the Ras-RafMEK-ERK signal transduction pathway *J. Cell Sci.* **122** 425–35

Skinnet S O, Xu H, Negarkar-Jaiwal S, Freire P R, Zwaka T P and Golding I 2016 Single-cell analysis of transcription kinetics across the cell cycle *eLife* **5** e12175

Smolen P, Baxter D A and Byrne J H 2000 Mathematical modelling of gene networks *Neuron* **26** 567–80

Stites E C, Aziz M, Creamer M S, Von Hoff D D, Posner R G and Hlavacek W S 2015 Use of mechanistic models to integrate and analyze multiple proteomic data sets *Biophys. J.* **108** 1819–29

Stites E C, Trampont P C, Haney L B, Walk S F and Ravichandran K S 2015 Cooperation between noncanonical ras network mutations *Cell. Rep.* **10** 307–16

Stites E C, Trampont P C, Ma Z and Ravichandran K S 2007 Network analysis of oncogenic Ras activation in cancer *Science* **318** 463–8

Stojilkovic S S, Tabak J and Bertram R 2010 Ion channels and signaling in the pituitary gland *Endocr. Rev.* **31** 845–915

Tian D and Kreeger P 2014 Analysis of the quantitative balance between insulin-like growth factor (IGF)-1 ligand, receptor, and binding protein levels to predict cell sensitivity and therapeutic efficacy *BMC Syst. Biol.* **8** 98–112

Tsaneva-Atanasova K, Sherman A, van Goor F and Stojilkovic S S 2007 Mechanisms of spontaneous and receptor-controlled electrical activity in pituitary somatotrophs: experiment and theory *J. Neurophysiol.* **98** 131–44

Uchida H, anba S, Wada M, Marsumoto K, Ikeda M, Naito N, Tanaka E and Honjo M 1999 Analysis of binding properties between 20 kDa human growth hormone (hGH) and hGH receptor (hGHR): the binding affinity for hGHR extracellular domain and mode of receptor dimerization *J. Mol. Endocr.* **23** 347–53

Veldhius J D, Carlson M L and Johnson M L 1987 The pituitary gland secretes in bursts: Appraising the nature of gland secretory impulses by simultaneous multiple-parameter deconvolution of plasma hormone concentrations *Proc. Natl Acad. Sci. USA* **84** 7686–90

Vilar J M, Guet C C and Leibler S 2003 Modeling network dynamics: the *lac* operon, a case study *J. Cell Biol.* **161** 471–6

Wagner C, Caplan S R and Tannenbaum G S 1998 Genesis of the ultradian rhythm of GH secretion: a new unifying experimental observation in rats *J. Physiol. (Endocrin. Metab.)* **275** E1046–54

Wan C *et al* 2019 LTMG: a novel statistical modeling of transcriptional expression states in single-cell RNA-Seq data *Nucleic Acids Res.* **47** e111

Wang R S, Saadatpour A and Albert R 2012 Boolean modeling in systems biology: an overview of methodology and applications *Phys. Biol.* **9** 055001

Wells J A 1996 Binding in the growth hormone receptor complex *Proc. Natl Acad. Sci. USA* **93** 1–6

Yang J, Bruno W J, Hlavacek W S and Pearson J E 2006 On imposing detailed balance in complex reaction mechanisms *Biophys. J.* **91** 1136–41

Yildirim N and Mackey M C 2003 Feedback regulation in the lactose operon: a mathematical modeling study and comparison with experimental data *Biophys. J.* **84** 2841–51

Zhou L and Chiu S Y 2001 Computer model for action potential propagation through branch point in myelinated nerves *J. Neurophysiol.* **85** 197–210

Chapter 6

Growth and remodelling

> Biological systems are so complex … that exact solutions are neither practicable nor useful.
>
> — C Thron

Continuum growth and remodelling models of biological tissues including the kinematic, constrained mixture and homogenised constrained mixture models are presented. The advantages and disadvantages of each approach are discussed.

6.1 Biological preliminaries

Cells of the human urinary bladder (UB) have a great degree of variability in their constitutive rates of turnover. For example: the renewal of the urothelium takes 3–6 months, although their regenerative capabilities are reduced to days following chitosan induced desquamation; the detrusor when subjected to repetitive stretching or cyclic intravesicular pressure changes shows a high rate of proliferation and differentiation (Khandelwal *et al* 2009, Birder *et al* 2010, Aitken and Bägli 2009, Heise *et al* 2012, Dai *et al* 2015, Fusco *et al* 2018). These responses are attributed to the action of GH, IGF-1, fibroblasts, adhesive and signal transduction molecules, cytoskeleton and ECM proteins, and inflammatory factors among others (Backhaus *et al* 2002, Bassuk *et al* 2003, Veranic *et al* 2009, Kreft *et al* 2005, Yang *et al* 2008, Wu *et al* 2012, Liang *et al* 2017, Barbosa *et al* 2017). The regeneration (apoptosis) varies with age and pathological conditions such as trauma, surgery and UB outlet obstruction (UBOO).

From this point onward, the main focus of investigation falls on mathematical modelling and simulation of bladder tissue growth and remodelling (G&R) related to UBOO. These involve the dynamic interplay and progressive cellular changes in the organ *per se* and extramural regulatory centres in the brain. Emphasis is given to

doi:10.1088/978-0-7503-3567-6ch6

the study of mechanical stimuli, extra- and intracellular signalling pathways and the endocrine mechanisms involved.

Histomorphologically, G&R of the human UB during outflow obstruction evolves in three sequential stages: hypertrophy, compensation and decompensation (Fusco *et al* 2018). During hypertrophy, changes in sensory transduction processes dominate in the urothelium. There exists a discernible disbalance between excitatory and inhibitory signalling, i.e. a reduced expression of μ_3-ACh type receptors and NO synthase along with an increased activity of purinergic $P2X_3$ and β_3-adrenoceptors. The detrusor enlarges in size and mass with an apparent widening of intercellular spaces among SMCs but without any evidence of hyperplasia. An insignificant amount of collagen and elastin is deposited within muscle bundles, inter- and intrafasciculi (Elbadawi *et al* 1993, Rubinstein *et al* 2007, Belucci *et al* 2017, Averbeck *et al* 2017, Mamoulakis and Oelke 2018). At molecular level, phosphoinositide 3 kinase regulated kinase, Ras/MAPK/ERK-1/2 and matrix metalloproteinases signalling pathways are activated.

The characteristic feature of the compensatory state is a continuous increase in synthesis and build-up of the extracellular ground matrix alongside a decrease in excitability and contractility of the detrusor.

With the continuous persistent obstruction, inflammatory and hypoxic mechanisms become more prominent. These are coupled with an increase in expression of inflammatory genes, the activation in production of RANTES, $p27^{Kip1}$, IL-6, HIF-1α, chemokines, and ERK-1/2 and NFκB pathways. A loss of smooth muscle cells, neuronal damage and fibrosis are apparent.

6.2 Continuum mechanics growth and remodelling models

6.2.1 General postulates

Perpetual morphostructural changes in biological tissues affect their integrative mechanical properties. The 19th century 'mechanistic' concept of interpreting life matter and the adaptive responses of living tissues to the dynamics of the environment (Davis 1867, Nutt 1913) underwent a revival in the 1970s emerging as a new and rapidly expanding trend in science dubbed mechanobiology. This aims to study the interactions and effects of mechanical stimuli on biological processes in cells and tissues, e.g. elongation/shortening of skeletal muscle in response to sustained stretching, the gyrification of the brain subject to internal stresses in a constrained volume of growth, cardiac ventricular dilatation and hypertrophy resulting from a chronic increase in blood pressure and volume overload, and adaptive morphofunctional changes in the UB caused by an increase in intravesicular pressure, among others. How to translate the gain and loss of mass associated with G&R into non-uniform changes in shape is a central question for mechanobiology.

The following paragraphs provide a brief overview of the general concepts and problems related to the modelling of G&R processes. Mathematical and biological advantages and limitations of models continue to be a subject for considerable scientific debate and are beyond the scope of this book. The reader interested in the subject is advised to consult specific sources (Ambrosi *et al* 2019, Latorre and

Humphrey 2018, Cyron and Humphrey 2017, Ateshian and Humphrey 2012, Ambrosi *et al* 2011).

6.2.2 Kinematic growth model

The fundamentals of kinematic growth in terms of modern continuum mechanics were formulated initially by Skalak *et al* (1982) and later elaborated and expanded by Rodriguez *et al* (1994) to become the general 'kinematic growth theory' of biocomposites. The cornerstone hypotheses adopted in the theory are that:

- the tissue is a homogenous biomaterial
- growth and remodelling proceeds via a series of stress-free fictitious intermediate configurations
- deformations of the tissue are finite with separability of the total deformation gradient by the multiplicative decomposition: $F = \mathbf{F}_e\mathbf{F}_g$ where $\mathbf{F}_e$, $\mathbf{F}_g$ are elastic and inelastic growth-related parts of the deformation, respectively

Assuming the tissue is incompressible under elastic deformation then: $\det(\mathbf{F}_e) = 1 \Rightarrow \det(\mathbf{F}) = \det(\mathbf{F}_e)\det(\mathbf{F}_g) = \det(\mathbf{F}_g)$. The mass density $\rho(t)$ in the current state is given by

$$\rho(t) = \rho_0(t)/\det(\mathbf{F}) = \rho_0(t)/\det(\mathbf{F}_g) \tag{6.1}$$

where $\rho_0(t)$ is the mass density in the referral configuration (see appendix D). If it is further assumed that there is a homogenous deposition of newly produced constituents within the tissue, then

$$\rho_0(t) = \det(\mathbf{F}_g)\rho(0) \quad \text{and} \quad d\rho_0(t)/dt = \rho(0)\frac{d\left[\det(\mathbf{F}_g)\right]}{dt}. \tag{6.2}$$

Equation (6.2) together with the constitutive choice for the Cauchy stress tensor

$$\boldsymbol{\sigma} = \rho\mathbf{F}\frac{\partial W}{\partial \mathbf{E}}\mathbf{F}^T, \tag{6.3}$$

the linear momentum equation

$$\text{div}(\boldsymbol{\sigma}) + \rho\mathbf{f}_b = 0, \tag{6.4}$$

the given evolution equation for $\mathbf{F}_g$, and the strain energy function per unit mass, W, comprise the mathematical problem for kinematic growth. In the above, ρ is the overall mass density, $\mathbf{f}_b$ is the body force.

The challenge in the above formulation is the construction of $\mathbf{F}_g$. Since it is not governed by any balance law, an appropriate constitutive equation, i.e. an evolution law for growth, must be postulated relating $\mathbf{F}_g$ to physical and chemical fields and biological signalling. When considered as a process of mass addition and loss, the equation of growth is written in terms of $\rho(t)$. However, a key question remains: how can this be related to the growth tensor? Two approaches are usually exercised:

the phenomenological

$$d\mathbf{F}_g/dt = \boldsymbol{\kappa}: \Delta\mathbf{G}, \tag{6.5}$$

and the 'structure' related type

$$\mathbf{F}_g = \sum_{j=1}^{m} \beta^j(t)\mathbf{B}^j. \tag{6.6}$$

Here $\boldsymbol{\kappa}$ is the 'gain-type' tensor, $\Delta\mathbf{G}$ is the growth stimulus, $\mathbf{B}^j$ are preferred growth directions, and $\beta^j(t)$ is scalar. The quantities $\boldsymbol{\kappa}$, $\Delta\mathbf{G}$, $\mathbf{B}^j$ and $\beta^j(t)$ appearing in (6.5) and (6.3) mimic phenomenologically a particular pattern of growth but do not reflect the complexity of the biological tissue under consideration or the nature of the intracellular processes engaged. Biological tissues are multiconstituent, i.e. they may contain smooth muscle, collagen and elastin fibres, epithelial tissue, nerve cells, etc. Each possess different mechanical characteristics and show dissimilar dynamics in G&R. The kinematic growth theory *per se* does not capture the individual responses of constructive morphological elements during the process of G&R and is thereby limited in its pragmatism.

6.2.3 Constrained mixture theory

The issue of the biological *naiveté* of the kinematic growth theory has been addressed in the constrained mixture G&R theory. Holzapfel *et al* (2000) constructed a mixture model (collagen and elastin fibres embedded in the ground substance matrix) of the tissues of large arteries. The approach was subsequently advanced by Humphrey and Rajagopal (2002, 2004) to exploit the basic principles of mixture and multi-network theories (figure 6.1). This allowed the modelling and study of mechanobiological processes 'in the biomaterial' along with the natural configurations of morphostructural constituents during their G&R (Rajagopal and Srinivasa 1998).

The basic constructive assumptions of the constrained mixture theory are that:

i. the biomaterial (biocomposite) is a multiphase system, a mixture of morphological structural elements (SMCs, collagen and elastin fibres, etc)
ii. individual constituents (superscript $i = 1, \ldots, n$) possess distinct mechanical characteristics
iii. deformations are finite and all morphostructural elements (superscript $i = 1, \ldots, n$) of the material deform together under externally/internally applied loads
iv. the total mass density of the tissue, ρ, and its constituents, ρ^i, in the current configuration remain constant; therefore: $\det(\mathbf{F}) = \sum_{i=1}^{n}\rho^i / \sum_{i=1}^{n}\rho_0^i$
v. simultaneous interstitial (volumetric) growth and deformation of constituents are a sequence of two mappings namely elastic, $\mathbf{F}_e^i$, and inelastic, $\mathbf{F}_g^i$, deformation; the inelastic $\mathbf{F}_g^i$ captures permanent changes of the local stress-free configuration caused by G&R whilst elastic $\mathbf{F}_e^i$ ensures the geometric continuity and mechanical equilibrium of the material, implying that the growth tensor effectively represents the addition (subtraction) of mass to a local volume element whilst the elastic tensor generates mechanical stresses

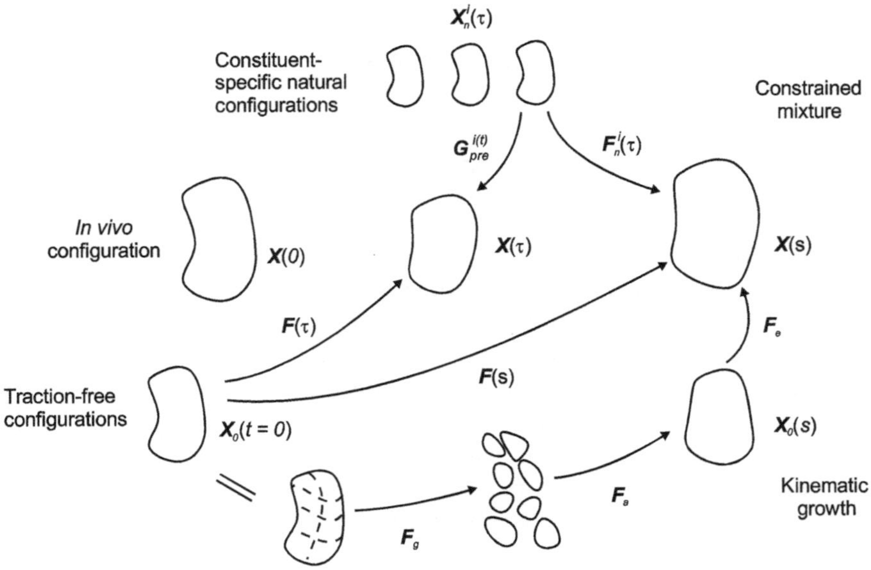

Figure 6.1. Schematic diagram of evolution in both the theory of kinematic growth and the constrained mixture theory. Reprinted from Ambrosi *et al* (2019) with permission of The Royal Society (UK).

vi. the elastic potential (strain energy per unit reference volume), Ψ is the sum of the strain energies of existing constituents, Ψ^i, such that: $\Psi^i = \rho_0^i W^i(\mathbf{F}_e^i)$; the use of the strain energy per unit volume is preferable since the mass changes considerably over the process of G&R

vii. the Cauchy stress in a constituent, $\boldsymbol{\sigma}^i$, is defined as

$$\boldsymbol{\sigma}^i = \frac{1}{\det(\mathbf{F})}\mathbf{F}\frac{\partial \rho_0^i W^i}{\partial \mathbf{E}}\mathbf{F}^T, \tag{6.7}$$

and the total Cauchy stress, $\boldsymbol{\sigma}$, in a contractile biocomposite is

$$\boldsymbol{\sigma} = \sum_{i=1}^{n} \frac{\rho\rho_0^i}{\rho_0\rho^i}\boldsymbol{\sigma}^i + \boldsymbol{\sigma}^{act}, \tag{6.8}$$

where $\sum_{i=1}^{n}(\rho\rho_0^i/\rho_0\rho^i)$ is the volume fraction of the ith constituent and $\boldsymbol{\sigma}^{\text{act}}$ is the active stress induced by SMCs

viii. since G&R is a slow ongoing process, the dynamic effect of inertia and volume forces can be neglected and, as such, the balance of momentum reduced to the quasistatic form

$$\text{div}\boldsymbol{\sigma} = 0 \tag{6.9}$$

ix. the evolution equation for $\mathbf{F}_g^i$ and ρ_0^i both being functions of mechanical (stress and strain) and biological (cell proliferation, arrangement, apoptosis, etc) processes are specified

x. true mass deposition, $d\rho^i_{0+}(t)/dt$, and degradation, $d\rho^i_{0-}(t)/dt$, occur at the rates defined by cell-specific processes
xi. new constituents are incorporated within extant tissue at homeostatic stretches.

The mass balance relation at time t of G&R is expressed as

$$d\rho^i_0(t)/dt = d(\rho^i_{0+}(t) - \rho^i_{0-}(t))/dt = \\ = \rho^i_0(t)Q^i(t) + \int_0^t \frac{d\rho^i_0(\tau)}{d\tau} q^i(t, \tau)d\tau, \quad \rho_0(t) = \sum_{i=1}^n \rho^i_0(t) \tag{6.10}$$

where $Q^i(t) \in [0, 1]$, $q^i(t, \tau) \in [0, 1]$ are survival functions, i.e. fractions of the mass of constituents produced at $\tau \in [0, t]$ surviving at current time t, and commonly chosen in the form

$$Q^i(t) = \exp(-t/T^i) \\ q^i(t, \tau) = \exp(-(t - \tau)/T^i). \tag{6.11}$$

In the above T^i is the average turnover time.

The strain energy function

$$W(\mathbf{F}) = \sum_{i=1}^n W^i(\mathbf{F}) \tag{6.12}$$

and $W^i(\mathbf{F})$ has the form

$$W^i(\mathbf{F}) = \frac{\rho^i_0(0)}{\rho_0(t)} Q^i(t) W^i(\mathbf{F}_e^{i(0)}) + \int_0^t \frac{1}{\rho_0(t)} \frac{d\rho^i_{0+}(\tau)}{d\tau} q^i(t, \tau) W^i(\mathbf{F}_e^{i(\tau)})d\tau. \tag{6.13}$$

The elastic deformation gradient $\mathbf{F}_e^{i(\tau)}$ of mass deposited at time τ is

$$\mathbf{F}_e^{i(\tau)} = \mathbf{F}(t) \cdot \mathbf{F}^{-1}(\tau) \cdot \mathbf{G}_{pre}^{i(\tau)}, \tag{6.14}$$

where $\mathbf{G}_{pre}^{i(\tau)}$ is the deposition stretch relative to the stress-free configuration of the mass increment, and $\mathbf{F}(t)$, $\mathbf{F}(\tau)$ are deformation gradients. By specifying $d\rho^i_0/dt$, e.g. $d\rho^i_0/dt = \rho^i_0 \boldsymbol{\kappa}^i_\sigma : (\boldsymbol{\sigma}^i - \boldsymbol{\sigma}^i_h)$, where $\boldsymbol{\sigma}^i_h$ is the homeostatic stress, i.e. the state during which there is no growth and remodelling, equations (6.7)–(6.14) present a closed problem for the biocomposite G&R.

The constrained mixture theory has gained popularity quickly as it captures mechanical changes in tissue resulting from unequal individual G&R of constituents. However, while this approach is biologically more plausible when compared to the kinematic model, it is computationally demanding. It requires the evaluation of the natural configurations of each constituent over the period of G&R. To overcome this difficulty, Watton *et al* (2004) proposed a heuristic approach based on rate equations to construct constrained mixture models. Mathematically it is a system of nonlinear differential equations which can be solved numerically both effectively and efficiently.

6.2.4 A homogenised constrained mixture model

More recently, advances on the constrained mixture models have been put forward by Cyron *et al* (2016). This latest approach seamlessly integrates micro-mechanical foundations with the structural simplicity of the earlier methods by providing a temporal homogenisation of effective mechanical parameters and state variables across all mass increments within a constituent. Following on from the general postulates (i)–(xi) of the constrained mixture theory, it is presumed that:

the deformation gradient yields (figure 6.2)

$$\mathbf{F} = \mathbf{F}_e^i \cdot \mathbf{F}_r^i \cdot \mathbf{F}_g^i, \tag{6.15}$$

where $\mathbf{F}_e^i$, $\mathbf{F}_g^i$ are elastic and inelastic deformations, respectively, and $\mathbf{F}_r^i$ describes remodelling, i.e. a change of the traction-free configuration of the ith constituent remodelling is given *a priori* by

$$\left[\frac{\partial \boldsymbol{\sigma}^i}{\partial \mathbf{F}_e^i} : \dot{\mathbf{F}}_e^i\right]_{\mathbf{F},\mathbf{F}_g^i=\text{const}} = -\left[\frac{\dot{\rho}_0^i}{\rho_0^i} + \frac{1}{T^i}\right][\boldsymbol{\sigma}^i - \boldsymbol{\sigma}_h^i], \tag{6.16}$$

From the above postulates it is possible to deduce: $\mathbf{F}_e^i = \mathbf{F} \cdot (\mathbf{F}_r^{-1} \cdot \mathbf{F}_g^i)^{-1}$, and thus to compute the strain energy function $W = \sum_{i=1}^{n} \rho_0^i W^i(\mathbf{F}_e^i)$, provided $\mathbf{F}_g^i$ is specified. When compared with the constrained mixture model (Humphrey and Rajagopal 2002), this approach has the computational advantage in that, during the process of its numerical realisation, only the tracking of the current mass density and inelastic deformation for each constituent are required.

6.2.4.1 A homogenised constrained mixture model of the bladder G&R

Homogenised constrained mixture models are suitable to model the G&R of thin-walled organs, e.g. arteries, the stomach, the urinary bladder, etc. Their growth manifests mainly as a change in thickness rather than in-plane enlargement of the

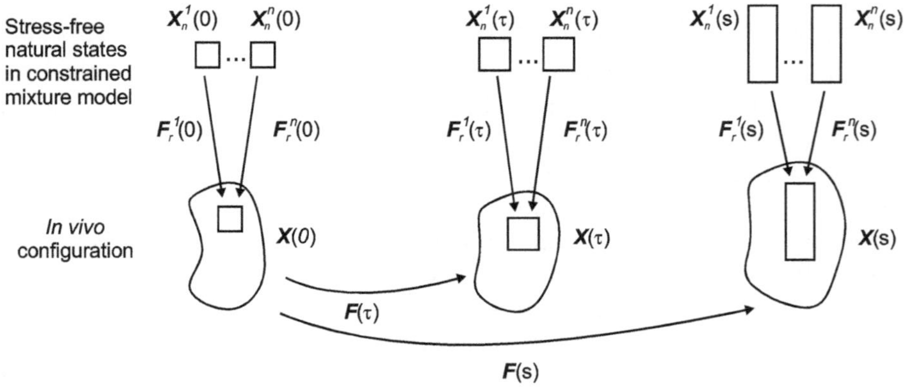

Figure 6.2. Schematic diagram of the homogenised constrained mixture theory. Reproduced from Cyron *et al* (2016) with permission from Springer.

wall (Braeu *et al* 2019). In the case of the UB, assume that the ground substance matrix is not subjected to G&R and the main changes occur only in the connective tissue and smooth muscle fibres (Holzapfel *et al* 2000, Baek *et al* 2006, Wilson *et al* 2012, Cyron and Humphrey 2014, Eriksson *et al* 2014, Grytsan *et al* 2015, Cyron *et al* 2016, Braeu *et al* 2019). The strain energy function for such fibres can be chosen in the form

$$W^i(t) = \frac{c_1^i}{2c_2^i}[\exp(c_2^i(\lambda_e^i)^2) - 1], \tag{6.17}$$

where c_1^i, c_2^i are material parameters and λ_e^i is the elastic stretch ratio in the direction $\mathbf{A}^i$ which corresponds to the alignment of the fibre family i and is calculated from

$$\begin{aligned} &\lambda_e^i = \mathbf{A}^i\!: \mathbf{F}_e^i \quad \text{or} \\ &(\lambda_e^i)^2 = (\mathbf{A}^i \otimes \mathbf{A}^i)\!: (\mathbf{F}_e^{i\mathrm{T}}\mathbf{F}_e^i) = (\mathbf{A}^i \otimes \mathbf{A}^i)\!: \mathbf{C}_e^i\,. \end{aligned} \tag{6.18}$$

Here $\mathbf{C}_e^i$ is the elastic Cauchy–Green deformation tensor ($\mathbf{C}_e^i = \mathbf{F}_e^{i\mathrm{T}}\mathbf{F}_e^i$), and the superscript T indicates the transposition. It follows from (6.18) that

$$\mathbf{F}_e^i = \lambda_e^i(t)\mathbf{A}^i \otimes \mathbf{A}^i. \tag{6.19}$$

Similarly, interpreting remodelling as an inelastic deformation gives

$$\mathbf{F}_r^i(t) = \lambda_r^i(t)\mathbf{A}^i \otimes \mathbf{A}^i \tag{6.20}$$

where $\lambda_r^i(t)$ is the inelastic remodelling stretch of the connective and SM fibres at a particular time t. From equations (6.15), (6.18)–(6.20) the total stretch of the fibres is deduced as

$$\lambda^i = (\mathbf{A}^i \otimes \mathbf{A}^i)\!: \mathbf{F} = \lambda_e^i\lambda_r^i. \tag{6.21}$$

Again, the total fibre stretch is decomposed multiplicatively into an elastic λ_e^i and inelastic, remodelling, part λ_r^i. Equation (6.21) does not include $\mathbf{F}_g^i$ since fibre growth affects the thickness but not the in-plane dimensions of the wall.

The evolution of λ_r^i satisfies (Cyron *et al* 2016)

$$\frac{d\lambda_r^i}{dt} = \left[\frac{d\rho_0^i(t)/dt}{\rho_0^i(t)} + \frac{1}{T^i}\right]\frac{\lambda_r^i}{\lambda_e^i}\frac{1}{\rho_0^i}\left[\frac{\partial W^i}{\partial \lambda_e^i} + \lambda_e^i\frac{\partial^2 W^i}{\partial^2 \lambda_e^i}\right]^{-1}(\sigma^i(t) - \sigma_h^i(t)) \tag{6.22}$$

where $\sigma^i = \boldsymbol{\sigma}^i\!: (\mathbf{A}^i \otimes \mathbf{A}^i)$, and T^i is the average turnover time. It reflects micro-structural alterations in the tissue dynamics to attain the physiological condition: $\sigma^i(t) = \sigma_h^i(t)$, where $\sigma_h^i(t)$ is the homeostatic fibre stress, i.e. an experimentally determined parameter specific to a tissue. The change of tissue mass and volume as a result of growth is given by (Cyron *et al* 2016)

$$\frac{d\rho_0^i(t)}{dt} = \rho_0^i(0)\kappa_\sigma^i\frac{\sigma^i(t) - \sigma_h^i(t)}{\sigma_h^i(t)}. \tag{6.23}$$

Here κ_σ^i defines the sensitivity of the biomaterial to deviations from homeostatic stress, whilst the meaning of the other parameters is as described above. Equations (6.22) and (6.23) stipulate the zero change of tissue mass if the Cauchy stress equals its homeostatic value σ_h^i and assumes its linear growth when $\sigma^i(t) > \sigma_h^i(t)$. It is rational to assume that mass growth is proportional to the existent mass since it is a direct result of cellular activity. Recalling assumption (vi), it follows that to describe the process of bladder growth *per se* the evolution equation (6.23) suffice.

It can be noted that if the bladder is modelled as a soft bioshell, growth is associated with a mere thickening of the wall (Wilson *et al* 2012, Wilson and Humphrey 2014). The mechanobiological reasoning behind this notion rests on there being significant differences between in-plane and cross-sectional stiffness. Therefore, for all constituents (Cyron and Humphrey 2017)

$$\mathbf{F}_g^i(t) = \left(\frac{\sum_i \rho_0^i(t)}{\sum_i \rho_0^i(t = 0)} - 1 \right) (\mathbf{F}\mathbf{A}_\perp) \otimes \mathbf{A}_\perp + I \tag{6.24}$$

where $\mathbf{A}_\perp$ is the normal direction of the surface wall in the reference configuration. Equation (6.24) becomes important only in problems related to the analysis of stress–strain distribution in the UB when the latter is modelled as a bioshell of inconstant variable thickness.

References

Aitken K J and Bägli D J 2009 The bladder extracellular matrix. Part I: Architecture, development and disease *Nat. Rev. Urol.* **6** 596–611

Ambrosi D *et al* 2011 Perspectives on biological growth and remodeling *J. Mech. Phys. Sol.* **59** 863–83

Ambrosi D, Mar M B, Cyron C J, DeSimone A, Goriely A, Humphrey J D and Kuhl E 2019 Growth and remodelling of living tissues: perspectives, challenges and opportunities *J. R. Soc. Interface* **16** 20190233

Ateshian G A and Humphrey J D 2012 Continuum mixture models of biological growth and remodeling: past successes and future opportunities *Ann. Rev. Biomed. Eng.* **14** 97–111

Averbeck M A, De Lima N G, Motta G A, Beltrao L F, Abboud Filho N J, Rigotti C P, Dos Santos W N, Dos Santos S K J, Da Silva L F B and Rhoden E L 2017 Collagen content in the bladder of men with LUTS undergoing open prostatectomy: a pilot study *Neurol. Urodyn.* **37** 1088–94

Backhaus B O, Kaefer M, Haberstroh K M, Hile K, Nagatomi J, Rink R C, Cain M P, Casale A and Bizios R 2002 Alterations in the molecular determinants of bladder compliance at hydrostatic pressures less than 40 cm H_2O *J. Urol.* **168** 2600–4

Baek S, Rajagopal K R and Humphrey J D 2006 A theoretical model of enlarging intracranial fusiform aneurysms *J. Biomech. Eng.* **128** 142–9

Barbosa J A B A, Reis S T, Nunes M, Ferreira Y A, Leite K R, Nahas W C, Srougi M and Antunes A A 2017 The obstructive bladder: expression of collagen, matrix metalloproteinases, muscarinic receptors, and angiogenic and neurotrophic factors in patients with benign prostatic hyperplasia *Urology* **106** 167–72

Bassuk J A *et al* 2003 Induction of urothelial cell proliferation by fibroblast growth factor-7 in RAG1-deficient mice *Adv. Exp. Med. Biol.* **539** 623–33

Belucci C H S, Riberio W O, Hemerly T S, de Bessa J Jr, Antunes A A, Leite K R M, Bruschini H, Srougi M and Gomes C M 2017 Increased detrusor collagen is associated with detrusor overactivity and decreased bladder compliance in men with benign prostatic obstruction *Prostate Int.* **5** 70–4

Birder L A, Wolf-Johnston A S, Chib M K, Buffington C A, Roppolo J R and Hanna-Mitchell A T 2010 Beyond neurons: Involvement of urothelial and glial cells in bladder function *Neurourol. Urodyn.* **29** 88–96

Braeu F A, Aydin R C and Cyron C J 2019 Anisotropic stiffness and tensional homeostasis induce a natural anisotropy of volumetric growth and remodeling in soft biological tissues *Biomech. Mod. Mechanobiol.* **18** 327–45

Cyron C J, Aydin R C and Humphrey J D 2016 A homogenized constrained mixture (and mechanical analog) model for growth and remodeling of soft tissue *Biomech. Mod. Mechanobiol.* **15** 1389–403

Cyron C J and Humphrey J D 2014 Vascular homeostasis and the concept of mechanobiological stability *Inter. J. Eng. Sci.* **85** 203–23

Cyron C J and Humphrey J D 2017 Growth and remodeling of load-bearing biological soft tissues *Meccanica* **52** 645–64

Dai Y, Tian Y, Luo D, Wazir R, Yue X, Li H and Wang K-J 2015 Cyclic stretch induces human bladder smooth muscle cell proliferation *in vitro* through muscarinic receptors *Mol. Med. Rep.* **11** 2292–98

Davis H G 1867 *Conservative Surgery* (New York: Appleton)

Elbadawi A, Yalla S V and Resnick N M 1993 Structural basis of geriatric voiding dysfunction. IV. Bladder outlet obstruction *J. Urol.* **150** 1681–95

Eriksson T S E, Watton P N, Luo X Y and Ventikos Y 2014 Modelling volumetric growth in a thick walled fibre reinforced artery *J. Mech. Phys. Sol.* **73** 134–50

Fusco F, Creta M, De Nunzio C, Iacovelli V, Mangiapia F, Li Marzi V and Agro E F 2018 Progressive bladder remodeling due to bladder outlet obstruction: a systematic review of morphological and molecular evidences in humans *BMC Urol.* **18** 15

Grytsan A, Watton P N and Holzapfel G A 2015 A thick-walled fluid–solid-growth model of abdominal aortic aneurysm evolution: Application to a patient-specific geometry *J. Biomech. Eng.* **137** 031008

Heise R, Parekh A, Joyce E M, Chancellor M B and Sacks M S 2012 Strain history and TGF-β1 induced urinary bladder wall smooth muscle remodeling and elastogenesis *Biomech. Mod. Mechanobiol.* **11** 131–45

Holzapfel G, Gasser T and Ogden R 2000 A new constitutive framework for arterial wall mechanics and a comparative study of material models *J. Elast.* **61** 1–48

Humphrey J D and Rajagopal K R 2002 A constrained mixture model for growth and remodeling of soft tissues *Math. Mod. Meth. Appl. Sci.* **12** 407–30

Humphrey J D and Rajagopal K R 2004 A constrained mixture model for arterial adaptations to a sustained step change in blood flow *Biomech. Model Mechano-biol.* **2** 109–26

Khandelwal P, Abraham S N and Apodaca G 2009 Cell biology and physiology of the uroepithelium *Am. J. Physiol. Renal Physiol.* **297** F1477–1501

Kreft M E, Sterle M, Veranic P and Jezernik K 2005 Urothelial injuries and the early wound healing response: tight junctions and urothelial cytodifferentiation *Histochem. Cell. Biol.* **123** 529–39

Latorre M and Humphrey J D 2018 Critical roles of time-scales in soft tissue growth and remodelling *APL Bioeng.* **2** 026108

Liang Z, Xin W, Qiang L, Xiang C, Bang-Hua L, Jin Y, De-Yi L, Hong L and Kun-Jie W 2017 Hydrostatic pressure and muscarinic receptors are involved in the release of inflammatory cytokines in human bladder smooth muscle cells *Neurourol. Urodyn.* **36** 1261–9

Mamoulakis C and Oelke M 2018 Diagnostic work-up of LUTS/BPH: From standard to new perspectives *Lower Urinary Tract Symptoms and Benign Prostatic Hyperplasia* (New York: Academic) ch 6 pp 113–33

Nutt J J 1913 *Diseases and Deformities of the Foot* (New York: EB Treat)

Rajagopal K and Srinivasa A 1998 Mechanics of the inelastic behavior of materials—Part 1, theoretical underpinnings *Int. J. Plast.* **14** 945–67

Rodriguez E K, Hoger A and McCulloch A D 1994 Stress-dependent finite growth in soft elastic tissues *J. Biomech.* **27** 455–68

Rubinstein M, Sampaio F J and Costa W S 2007 Stereological study of collagen and elastic system in the detrusor muscle of bladders from controls and patients with infravesical obstruction *Int. Braz. J. Urol.* **33** 33–9

Skalak R, Dasgupta G, Moss M, Otten E, Dullemeijer P and Vilmann H 1982 Analytical description of growth *J. Theor. Biol.* **94** 555–77

Veranic P, Erman A, Kerec-Kos M, Bogataj M, Mrhar A and Jezernik K 2009 Rapid differentiation of superficial urothelial cells after chitosan-induced desquamation *Histochem. Cell Biol.* **131** 129–39

Watton P, Hill N and Heil M 2004 A mathematical model for the growth of the abdominal aortic aneurysm *Biomech. Model. Mechanobiol.* **3** 98–113

Wilson J S, Baek S and Humphrey J D 2012 Importance of initial aortic properties on the evolving regional anisotropy, stiffness and wall thickness of human abdominal aortic aneurysms *J. Roy. Soc. Interf.* **9** 2047–58

Wilson J S and Humphrey J D 2014 Evolving anisotropy and degree of elastolytic insult in abdominal aortic aneurysms: potential clinical relevance? *J. Biomech.* **47** 2995–3002

Wu T, Chen L, Wei T, Wang Y, Xu F and Wang K 2012 Effect of cyclic hydrodynamic pressure-induced proliferation of human bladder smooth muscle through Ras-related C3 botulinum toxin substrate 1, mitogen-activated protein kinase kinase ½ and extracellular regulated protein kinases ½ *Int. J. Urol.* **19** 67–74

Yang R, Amir J, Liu H and Chaqour B 2008 Mechanical strain activates a program of genes functionally involved in paracrine signaling of angiogenesis *Physiol. Genom.* **36** 1–14

Roustem N Miftahof and Christian J Cyron

Chapter 7

Brain–bladder axis in tissue growth and remodelling

The pessimist complains about the wind; the optimist expects it to change; the realist adjusts the sails.

— William A Ward

A mathematical model of the multicompartmental urinary bladder–brain axis involved in growth and remodelling of the bladder wall tissue is constructed relying on mechanical and biological data of both structure and function of each compartment. Numerical simulation results of the bladder's response to outflow obstruction are presented. Problems of identification and validation of parameters and constants of the model are discussed.

7.1 The architecture of the BBA

The construction of a unified mathematical framework for the urinary bladder–brain axis (BBA), showing the endocrine regulation of tissue growth and remodelling in relation to UBOO, is a daunting task. Globally no attempts have been made as yet to simulate these processes within a closed self-controlled biomechanical system. The existing mathematical models of the hypothalamus-adenohypophysis relationship, as discussed in chapter 5, can replicate only the secretory patterns of SS, GH, GHRH, and IGF-1 in response to paracrine stimuli and do not consider: (i) associated cellular electrical phenomena, (ii) the co-release and co-transmission by multiple neurotransmitters and hormones, (iii) (the) receptor polymodality and co-localisation, or (iv) intracellular signalling pathways. Therefore, they remain basic, fragmentary and ill equipped to study the neurohormonal control of bodily tissue G&R.

doi:10.1088/978-0-7503-3567-6ch7

The BBA comprises multiple compartments including the urinary bladder *per se*, mechanosensory neurons, the DRG, the intermediate sensory neurons in the laminae I–II of the SDH in the spinal cord, neurons of the PeN and ArcN-ME in the hypothalamus, somatotrophs in the *pars anterior* of the pituitary gland, and hepatocytes in the liver (figure 7.1(a)–(c)). The postulates exploited in mathematical formulation of the model rely on the mechanical and biological data of the structure and function of each compartment and are summarised below:

(i) the urinary bladder is a thin soft biological shell; its wall is composed of two SM layers, collectively known as the detrusor, embedded in the ECM; SM fibres in the outer layer are orientated longitudinally and in the inner layer, circumferentially with respect to the neck-apex axis of the organ

(ii) the detrusor is an orthotropic viscoelastic continuum; the total force generated by the detrusor is the sum of elastic and active forces; these are produced by collagen and elastin (elastic), and electrically induced contractions by SMCs (active), respectively; the key player of the electromechanical coupling is Ca_i^{2+}

(iii) electrically, the detrusor is an isotropic syncytium; its excitability is provided by neurochemically modulated transmembrane voltage-dependent Ca^{2+} channels of L- and T-types, large conductance Ca^{2+}-activated K^+ channels, potential sensitive K^+ channels, and Cl^- ion channels

(iv) the existence of intramural pacemaker-like cells is presumed

(v) coordinated mechanical contractions are controlled by an intramural network of ganglia; each ganglion is formed of primary sensory, inter- and motor neurons

(vi) mechanoreceptors sense the deformation (stretching) of the bladder; this information is coded and transmitted as nerve ending and dendritic APs along the myelinated Aδ-type nerve fibres of the hypogastric nerve to the neuron in the lamina I/II of the sSDH

(vii) signal transduction at the soma of the SDH neuron involves L-Glu and GABA and cognate $NMDAR_1$, $GluR_{1/2}$ and $GABA_A$ receptors; the electrical response of the neuron is a result of the integrated activity of transmembrane Ca^{2+}, BK_{Ca}, K^+, Na^+, and Cl^- channels and related intracellular signalling pathways, e.g. sGC/PkG

(viii) the SDH neuron projects the unmyelinated C-type axon to the PeVN in the hypothalamus; these produce, store and release SS on demand; the processes are regulated neurochemically and hormonally by L-Glu, GABA, SS, GHRH, GH, IGF-1 and Ghr acting via cognate $NMDAR_1$, $GABA_A$, $mGluR_{1/2}$, $sst_{1/2}$, GHRH-R, and tyrosine kinase GH and IGF-1 receptors and multiple crosslinked intracellular signalling pathways, e.g. PLCγ/Ras/MAPK/ERK1-2, AC/cAMP/PkA, sGC/PkG, JAK-(in)dependent, etc, and transmembrane L-, T-type Ca^{2+}, Na^+, K_{ir}, BK_{Ca} ion channels; the PeV neuron has extensive connections to the ArcN-ME complex

(ix) the ArcN-ME complex neurons synthesise, store and release GHRH into the hypophesial portal blood system; the extrasynaptic exocytosis of the

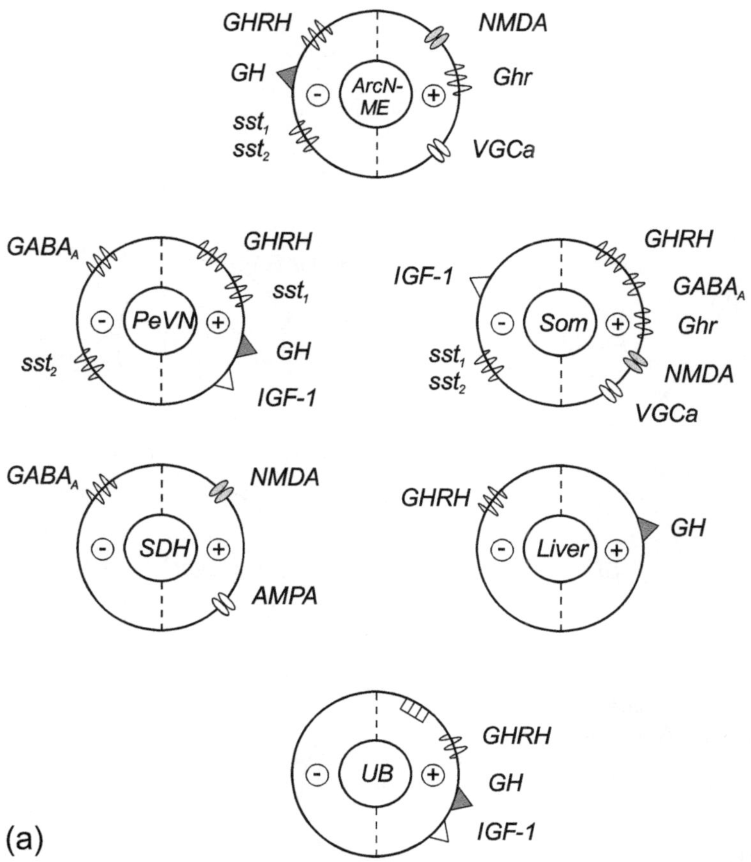

Figure 7.1. (a) Topographical distribution of receptors with cognate neurotransmitters and hormones in morphostructural elements of the BBA under consideration. Here: are GPCRs, where indices 1 through 5 indicate $GABA_A$, sst1, sst2, GHRH and Ghr receptors, respectively, , are ionotropic NMDA and AMPA receptors, , are GH and IGF-1 receptors, is the mechano-receptor, physiological effects produced, i.e. excitation and inhibition, as a result of receptor activation are indicated by inner circles marked + and −, respectively. (b) Extracellular signalling pathways among morphostructural elements of the BBA. Electrochemical connections with types of fibres and neurotransmitters involved are indicated by solid lines; hormonal links are shown by dotted lines.

hormone is upregulated electrically and neurohormonally by L-Glu, Ghr, and IGF-1 acting on related $NMDA_1$, $mGluR_1$, $GhrR_1$ and IGF-1R receptors; GHRH secretion is downregulated by GHRH and SS via the GHRH-R and $sst_{1/2}$ receptors, respectively; the intracellular mechanisms and pathways involved are similar to those described above (viii), (ix); it is presumed that Ghr can directly modulate the permeability of transmembrane voltage-dependent Ca^{2+} channels;

(x) GH is synthesised, stored and released by somatotrophs of the adenohypophysis; hormone exocytosis into the systemic blood circulation is upregulated by GHRH and Ghr acting on the cognate GHRH-R and

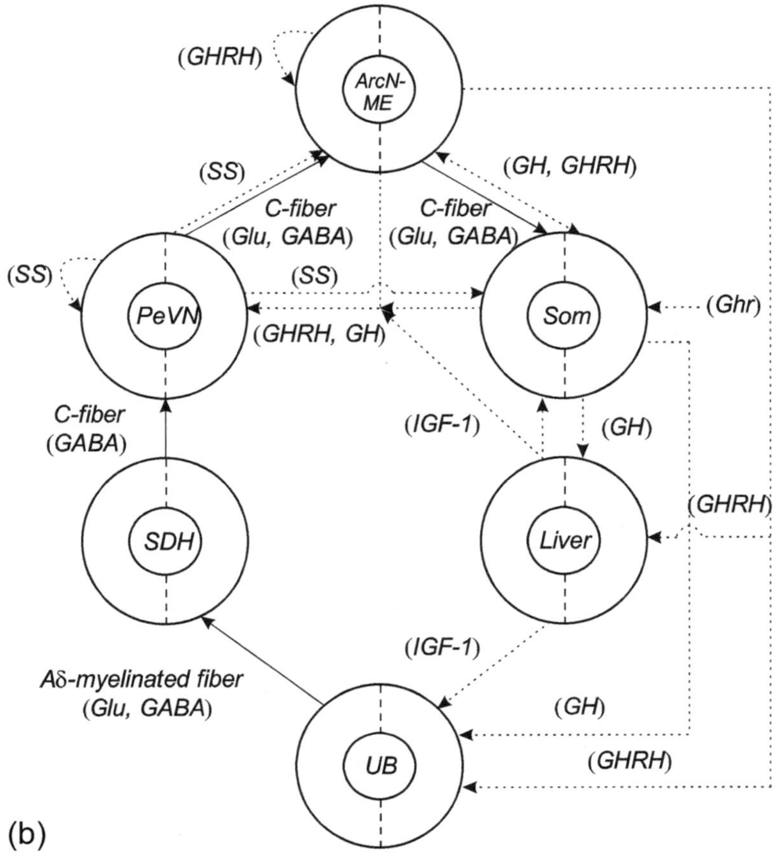

Figure 7.1. (Continued.)

GhrR$_1$ receptors coupled with AC/cAMP/PkA and PLC/PIP$_2$/PkG signalling pathways; downregulation of GH secretion and synthesis is under the control of SS and IGF-1; somatotrophs' self-oscillatory electrical patterns of discharges are the result of the dynamic interplay among the T- and L-type Ca^{2+}, K_{dr}, K_{ir}, BK_{Ca}, Na^+, and Cl^- transmembrane ion channels

(xi) production and exocytosis of IGF-1 is upregulated by GH and downregulated by GHRH

(xii) the hypertrophy, compensation and decompensation of SMCs, growth and remodelling of the collagen/elastin network are regulated by GH and IGF-1 acting on their cognate receptors coupled with the JAK-(in) dependent and PI3/PkB/S6k1/S6 signalling pathways; they induce gene expression of myogenesis and collagenesis leading to cell proliferation and degradation.

The justification and verification of anatomical and morphological arrangements, electrophysiological and biochemical intracellular signalling pathways are discussed in full detail in preceding chapters.

7.2 Mathematical formulation of the BBA

The model incorporates extensive details of morphofunctional components of the BBA, being multi-scale and large in terms of the numbers of subsystems and parameters represented. The challenge is to build a non-trivial framework that merges these subsystems with nominal uncertainty and bias to replicate realistically arranged biological processes. In the current model: the convective transport of hormones by blood, the effects of chemotaxis, the mechanisms of gene transcription, and cell proliferation, differentiation and rearrangement have not been simulated. Assuming that G&R is a mechanobiological homeomorphism, the combined system of equations comprising the closed mathematical problem for the BBA is:

(i) Equations (3.7), (3.13)—the urinary bladder biomechanical response to filling and voiding with the production of active, $T^{\mathrm{a}}_{c,l}$, and passive, $T^{\mathrm{p}}_{c,l}$, forces and deformations, $\lambda_{c,l}$

(ii) Equations (3.25), (3.22)–(3.24)—the propagation of the wave of depolarisation, V_{d}, within the detrusor syncytium

(iii) Equations (3.26)–(3.30)—the self-oscillatory electrical activity of the detrusor, V

(iv) Equations (3.14)–(3.18)—the 'pacemaker'-like electrical activity and the generation of V_{p} potentials

(v) Equations (3.31), (3.34)—the constitutive relations for the bladder wall tissue and electromechanical coupling

(vi) Equations (3.35)–(3.37)—the dynamics of mechanosensation with the production of nerve ending, V_{se}, and dendritic, V_{sd}, potentials

(vii) Equations (5.4), (5.5)—the propagation of the action potential, V_{MS}, along the myelinated Aδ fibre of the hypogastric nerve

(viii) Equations (4.1)–(4.3), (4.4)–(4.6)—the electrochemical signal transduction at the glutamatergic and GABAergic synapses, e.g. the Aδ fibre-soma of the SDH sensory neuron, the axo-somatic synapse on the PeVN neuron

(ix) the equation for the generation of EPSP/IPSP, $V^{(+/-)}_{\mathrm{syn}}$,

$$C_{\mathrm{p}}\frac{dV^{(+/-)}_{\mathrm{syn}}}{dt} + V_{\mathrm{syn}}\left(\mp\Omega[\ldots] + R_v^{-1}\right) = \frac{V_{\mathrm{syn},0}}{R_v}, \tag{7.1}$$

where C_{p} is the capacitance of the postsynaptic membrane, R_v is the resistance of the synaptic structures, Ω is the empirical constant and $V_{\mathrm{syn},0}$ is the resting postsynaptic potential, [...] is the concentration of the respective active neurotransmitter-receptor complex, i.e. x_{10},x_{11} and x_{19} (chapter 4)

(x) the production of the total (summatory) postsynaptic potential, $V^{\mathrm{sum}}_{\mathrm{syn}}$ on the soma of the SDH neuron

$$V^{\mathrm{sum}}_{\mathrm{syn}}(t) = V^{+}_{\mathrm{syn}}(t) + V^{-}_{\mathrm{syn}} \tag{7.2}$$

(xi) Equations (5.16)–(5.20)—the generation of APs (V_{sdh}, V_{PeV}, $V_{ArcN-ME}$) at the somas of the SDH, PeV and Arc-ME neurons

(xii) Equation (5.3)—the propagation of electrical signals along the unmyelinated fibres, e.g. the links between SDH and PeVN neurons, PeV and the ArcN-ME complex neurons
(xiii) Equations (4.8) and (4.9), (4.11) and (4.12), (4.14) and (4.15), (4.17) and (4.18), (4.20) and (4.21)—the separate or conjoint endocrine transduction by SS, GHRH, Ghr, GH and IGF-1, respectively
(xiv) Equations (4.23)–(4.24)—myogenesis
(xv) Equation (4.25)—collagenesis
(xvi) Equations (6.10), (6.12)–(6.18), (6.24)—continual G&R of the tissue.

The system, at $t = 0$, is at a mechanically and electrically quiescent state, i.e. all transmembrane potentials and ion channels are in relaxed, resting states. The concentrations of reacting components (neurotransmitters and hormones) are known and correspond to average physiological values, and the undeformed, non-stretched configuration of the UB is predefined by a given intravesicular pressure $p_0 = 1$ kPa.

Boundary conditions assume that: the neck of the urinary bladder is rigidly fixed; the proximal ends of (un)myelinated nerve fibres are excited by internally generated signals; the synaptic ends remain electrically unexcitable throughout. Excitation to the system is provided by the intraluminal pressure p.

7.3 An Achilles' heel

The mathematical model, as described above, relies on nearly 300 variables including: geometrical and electrical constants of neurons and cells, association/dissociation rates of biochemical reactions, and concentrations of ions, neurotransmitters and hormones. Their real values are unknown and the physiological range for most spans several orders of magnitude. This proves to be the Achilles' heel of the model. In addition to a parametric quantitative tuning of the system, its core topology also embodies admissible sets of solutions (which are) tolerant to intrinsic perturbations. Thus, a search for a particular solution that could be tested against *in vivo* and *in vitro* investigations is more rational rather than a search for a unique, globally optimal parameter set which, in this case, is biologically senseless.

The notion of using experimentally acquired parameters only provides coherence guaranteeing the proper calibration of models against the information available. However, this approach has the following serious drawbacks (Babtie and Stumpf 2017): (i) being abstractions of complex processes, the meaning of variables in a model has to be thoroughly gauged, e.g. the commonly employed Michaelis–Menten kinetics may grossly simplify the dynamics of enzymatic reactions and, hence, rates of chemical reactions may not reflect the actual measured biochemical constants; (ii) generally, parameters are obtained from *in vitro* assays or related species rather than from their natural *in vivo* context which may differ in orders of magnitude thus making the former measurements misleading; (iii) most parameter sets used in complex models are collated and annotated from heterogeneous sources in a nonsystematic, nonstandardized and inconsistent way; (iv) uncertainty and noise that habitually pervade biological systems are often not considered; (v) the restricted choice of variables measured solely experimentally incurs an experimentally limited bias in the model. Although it is desirable to use *in vivo* information from the target organism to help obtain relevant parameters, it is equally important to exploit other techniques and methods to assess the missing data.

Statistical inference has proven effective in the parameterisation of small-scale systems such as individual intracellular signalling pathways, regulatory motifs, metabolic reactions and receptor complexes (Zamora-Sillero *et al* 2011). To find the right parameter from a randomly chosen parameter set, θ, and, if possible, to assess explicitly its associated uncertainty using statistical methods, the likelihood, $L(\theta)$, or the probability, $\Pr(D|\theta)$, of observing the data $D = \{d_1, \dots, d_n\}$ is introduced (Cox 2006)

$$L(\theta) = \Pr(D|\theta) = \prod_{n}^{i=1} \Pr(d_i|\theta) \tag{7.3}$$

where d_i is an experimental observation for a mathematical model and chosen θ. A particular solution $\hat{\theta}$, the best parameter set, is obtained by varying θ until

$$\hat{\theta} = \arg\max(L(\theta)). \tag{7.4}$$

The likelihood function evaluated over the parameter space, Ω_θ, is complex and multimodal in nature relying heavily on numerical optimisation algorithms. Local, global and profile optimisation techniques have been exercised (Raue *et al* 2009, Fröhlich *et al* 2014, Sun *et al* 2012, Moles *et al* 2003). In the proposed BBA model, if it is assumed that the tolerated variation in the average parameter is 10%, statistical inference would find only one solution in 10 300 samples (!) for which simulations would exhibit the desired responses.

Compared to statistical inference which provides a single assessment of $\hat{\theta}$, the Bayesian inference offers the probability of posterior distribution over Ω_θ that reflects the level of existing (equally lacking) knowledge of the system. Thus, $L^p(\theta)$ combined with the prior probability $\pi(\theta)$ yields

$$L^p(\theta) = \Pr(\theta|D) = \frac{\Pr(D\,|\,\theta)\pi(\theta)}{\int_{\Omega_\theta} \Pr(D|\theta)\pi(\theta)d\theta}. \tag{7.5}$$

Although the method is advanced in its interpretational capabilities, its application in realistic problems is computationally unfeasible. Therefore, approximate Bayesian and other likelihood-free methods have been developed recently (Liepe *et al* 2014, Jagiedella *et al* 2017).

An alternative approach to the problem is the 'direct' calculation of kinetic constants of diffusion-limited reactions using the Brownian dynamics (BD) trajectory simulation method. This has shown its effectiveness in assessments of protein–ligand and protein–substrate association reactions. Assuming the proteins are rigid bodies, thus neglecting internal conformational changes and the flexibility of side-chains, their dynamics are subjected to a random diffusion Brownian process.

Ermak and McCammon (1978) proposed an algorithm for calculating phase-space trajectories and the distribution function $W(\{r_i\}, \{p_i\}, t)$—the average number density of particles at time t at a location $\{r_i\}$ and a momentum $\{p_i\}$—of an ensemble of proteins (particles) in solution. Making an assumption of rapid momentum relaxation, i.e. the momentum variables relax to equilibrium much faster than the position variables, the molecular motion is described by the modified Langevin equation

$$\sum_j \tau_{ij}\dot{\upsilon}_j = \upsilon_i + \frac{1}{kT}\sum_j D_{ij}F_j + \sum_j \sigma_{ij}f_j, \tag{7.6}$$

where $\tau_{ij} \equiv D_{ij}m_j/kT$, $\sigma_{ij} = \frac{1}{kT}\sum_j D_{ij}\alpha_{lj}$. The meaning of parameters related to a particle are: υ_i—velocity, F_j—the sum of interparticle and external forces acting in direction j, m_j—the mass, D_{ij}, α_{lj}—configuration-dependent diffusion and hydrodynamic friction tensors, respectively, k—Boltzman's constant, and T—the absolute temperature. The last term in equation (7.6) represents a randomly fluctuating force exerted on a protein by the solution with f_j defined by a Gaussian distribution. Expanding τ_{ij} in the Taylor series about the initial position of particles and integrating twice over time, the equation for diffusion, Δr_i, of a protein yields

$$\Delta r_i(t) = \sum_j\left(\frac{\partial D_{ij}^0}{\partial r_i}t + \frac{D_{ij}^0F_j^0}{kT}t + \sigma_{ij}^0\int_0^t [1 - \exp((s - t)/\tau_{ii}^0]f_j(s)ds\right), \tag{7.7}$$

from where the mean,

$$\langle\Delta r_i\rangle = \sum_j\left(\frac{\partial D_{ij}^0}{\partial r_i} + \frac{D_{ij}^0F_j^0}{kT}\right)t \tag{7.8}$$

and the covariance,

$$\langle\Delta r_i\Delta r_j\rangle \approx 2D_{ij}^0t \tag{7.9}$$

of the displacements can be assessed.

Northrup *et al* (1984) applied the above algorithm to estimate the rate association constants, k_{PL}, of bimolecular protein–ligand reactions

$$k_{PL} = \frac{k_D(b)\left[\frac{\beta}{1-(1-\beta)\Omega}\right]\alpha}{1 - (1 - \alpha)\left\{\Delta + \left[\frac{\beta}{1-(1-\beta)\Omega}\right](1 - \Delta)\right\}}. \tag{7.10}$$

Here: α, $\beta(r)$, $\Omega(r)$, $\Delta(r)$ are intrinsic collision, escape, recombination and recollide probabilities, respectively, for a particle at the reactive site; $k_D(b)$ is the rate constant at which particles achieve the *a priori* set distance $r = b$ whether they react subsequently or not. These are calculated by the BD trajectory method.

To account for the differences in the dielectric nature of protein and solvent hydrodynamic multiparticle interactions, a range of improved models have been suggested (Gabdoulline and Wade 2001, 2002, Geyer and Winter 2009, Huber and McCammon 2019). These have proved successful in studying the mechanisms and kinetics of DNA-chromatin, heterogenous protein mixtures, CDC42 GTPase binding, linker histones, the p53 tumour suppressor, the absorption of proteins to surfaces, PkA and the influenza virus (Pang and Zhou 2016, Hirakis *et al* 2017, Amaro *et al* 2018, McGuffee and Elcock 2010, Pachov *et al* 2011, Haddadian and Gross 2006, Ozturk *et al* 2018, ElSawy *et al* 2016, Romanowska *et al* 2015). However, the BD method does not allow the effective analysis of microscopic mechanisms, i.e. the protein–ligand association pathways, that lead to the actual formation of protein–ligand complexes.

The statistical physics transition path theory has been adapted more recently to describe and calculate the overall reaction constants. In addition to the general assumptions of the BD trajectory model, the conformation space, Ω_{PL}, is conceptually divided into dissociated and reacting zones connected via the absorbing boundary. The protein diffuses freely with a rate k_{Diff} in the former, and interacts with the ligand electrostatically and hydrodynamically at a rate, k_{PL}, in the latter. The overall association rate constant, k_+, is given by

$$k_+ = \frac{k_{Diff} k_{PL}}{k_{-Diff} + k_{PL}}. \tag{7.11}$$

It follows that if $k_{PL} \gg k_{-Diff}$, then $k_+^D \approx k_{Diff}$, i.e. the reaction is diffusion limited. Otherwise, if $k_{PL} \ll k_{-Diff}$, then $k_+^R \approx k_{PL} k_{Diff} / k_{-Diff}$, i.e. the reaction is rate limited. These evaluations provide a crude estimate for the range of variation of k_+: $n \cdot 10^{-3} \leqslant k_+ \leqslant n \cdot 10^3$ (μM-1s-1) (Sykes *et al* 2019, Schreiber *et al* 2009).

The k_{Diff} can be obtained either by solving one of the Smoluchowski, Langevin, Fokker–Planck equations or through Brownian dynamics simulations. For example, Smoluchowski (1917) obtained the explicit formula

$$k_+^D \approx 4\pi Dr, \tag{7.12}$$

where D is the relative transitional diffusion constant. Although it is simple and easy to use, it does not address the issue of orientational constraints with reference to the stereoscopic association of proteins and offers unrealistic results. Thus for proteins with the surface fractions $f_{1,2}$ covered by small reactive patches, Šolc and Stockmayer (1973) suggested calculating k_{Diff} as

$$k_+^D \approx 4\pi Dr[f_1 \xi_2 \tan(\delta_2/2) + f_2 \xi_1 \tan(\delta_1/2)], \tag{7.13}$$

where $\delta_{1,2}$ are patch spanning polar angles, and $\xi_i (i = 1,2)$ are given by

$$\xi_i = \sqrt{\frac{1}{2} + \frac{D_i r^2}{2D}}. \tag{7.14}$$

Here D_i is the rotational diffusion constant. The effect of centrosymmetric electrostatic interaction between proteins with reactive patches can be included in the expression derived by Zhou (1993, 1997)

$$k_+^D \approx \frac{4\pi Dr\exp(-U(r)/kT)[P_1(0)P_2(0)]}{\sum\limits_{x,x_1,x_2=0}^{\infty}\left[-\frac{\rho_x(r)/r}{d\rho_x(r)/dr}\right]_{r=R}[P_1(x_1)P_2(x_2)]^2P_{xx1/2}}, \tag{7.15}$$

where $U(r)$ is a Coulomb potential, $P_i(x_i)$ are Legendre polynomials, $P_{xx1/2}$ are defined by Wigner 3-j symbols, and $\rho_x(r)$ satisfies

$$\begin{aligned} &r^{-2}\exp(\beta U(r))\frac{d}{dr}(r^2\exp(-\beta U(r)))\frac{d\rho_x(r)}{dr} \\ &=\left[\frac{x_1(x_1+1)D_1+x_2(x_2+1)D_2}{D}+\frac{x(x+1)}{r_2}\right]\rho_x(r) \end{aligned} \tag{7.16}$$

with the condition: $\rho_x(r) \to 0$ as $r \to \infty$.

To calculate k_{PL}, the reacting zone is subdivided further into small subzones ($i = 1, \ldots n$). Transition rates $k_{i,i+1}$ (transition probabilities, $T_{i,i+1}(l)$, where l is the length of a trajectory) between the adjacent subzones i and $i+1$ are calculated separately using Markov models. Through reassembly, the entire dynamics of the pathway can be reconstituted. $k_{i,i+1}$, or equally $T_{i,i+1}(l)$, are used to both evaluate the probability distribution of reactive trajectories (the committor q_i), i.e. trajectories that finally progress to binding to the ligand, and to find the reaction rate k_{PL} (Noé *et al* 2009)

$$k_{PL} = \frac{\sum\limits_{i\in\Omega_{PL}}\sum\limits_{i\notin\Omega_{PL}}\pi_i k_{i,i+1}q_{i+1}}{\sum\limits_i \pi_i(1-q_i)}. \tag{7.17}$$

In the above, π_i is a stationary distribution of each subzone (the Boltzman distribution).

In vitro radioligand binding assays are the gold standard for obtaining the rate constants of biochemical reactions (Sykes *et al* 2019). An array of identical cognate receptors are labelled with increasing concentrations of a radioisotope in the presence and absence of an unlabelled competitive ligand (protein). The kinetic traces of ligand-receptor binding *vs* discrete moments of time are recorded and fitted to non-specific bimolecular or Motulsky and Mahan equations to derive the association constant k_+ (Motulsky and Mahan 1984). The dissociation constant k_- is obtained either by adding a high concentration of a competitor or by diluting 'infinitely' the tracer. Although the method offers the most direct and accurate measurement of $k_{+/-}$ for ligands with high affinity for a receptor, it does present a number of experimental challenges and intrinsic pitfalls. The most compelling evidence is that: (i) the radioactive probes being used are hazardous, (ii) the tracers cannot be read continuously, (iii) the binding assays cannot perform in a homogenous format since the bound and unbound fractions of radioligand have to be separated, (iv) the approach *per se* does not account for an allosteric ligand-receptor interaction, (v) it is impossible to discriminate between affinities for orthosteric and non-specific receptor binding sites. Bead-based scintillation assays overcome the problem of homogeneity despite suffering from bead settling at the beginning of recordings which may lead to inaccurate measurements at a later stage (Xia *et al* 2016).

The design and synthesis of high affinity fluorescent tracers for various receptors serve as a foundation for fluorescence-based methods, an alternative to radioactive assays. The fluorescent ligand-receptor association is measured using confocal microscopy despite the small signal-to-noise ratio hindering accuracy in assessing non-specific binding kinetics. This difficulty has been oversome by time resolved fluorescence and bioluminescence resonance energy transfer methods, TR-FRET and BRET, respectively. In the TR-FRET method SNAP, CLIP or ACP-tagged receptors are labelled with terbium cryptate which, upon laser excitation at a wave length ~ 340 nm, transfer energy to an acceptor fluorophore on a fluorescent ligand positioned at a distance < 10 nm (Schiele *et al* 2015). In the BRET method, small luciferase NanoLuc is used instead as a tag which, in the presence of its substrate furamizine, produces a bioluminescent light (Hall *et al* 2012). Once in close proximity, the energy emitted upon furamizine oxidation is transferred to the fluorophore of the acceptor. The recorded data, signal *vs* time, are employed to calculate rate constants using the Motulsky and Mahan equation.

The advantages of these methodologies are apparent: (i) the signal is specific to ligand-receptor binding, (ii) the assay format is homogenous, and (iii) the kinetic resolution is higher than in radioligand assays. The main limitation, however, is the possible alteration in receptor properties brought about by its tagging with an exogenous substrate. Nevertheless, these technologies have been successfully applied to study the receptor tyrosine kinase and vascular endothelial growth factor receptor 2 (Kilpatrick *et al* 2017, Peach *et al* 2018).

It is obvious that kinetic rates of the association/dissociation of proteins are contingent on multiple factors such as ligands': molecular structure, weight, the distribution and geometry of binding sites, conformational changes, translational and rotational diffusion, mutagenesis of surface residues, electro- and hydrostatic interactions, as well as the properties of the intra-extracellular environment including viscosity, ionic strength, crowding and depletion. Rather than being guided by any analytical model that only partially considers the aforementioned factors, $k_{+/-}$ are usually estimated on an ad hoc basis. The physiological range of $k_{+/-}$ for reaction-limited and diffusion-limited protein associations is: $n \cdot 10^{-3} \leqslant k_+^R \leqslant n \cdot 10^{-1}$ ($\mu M^{-1}s^{-1}$) and $n \cdot 10^{-1} \leqslant k_+^D \leqslant n \cdot 10^4$ ($\mu M^{-1}s^{-1}$), respectively. For proteins such as GH, IGF-1, GHRH, kinases etc, the k_+^D vary within $n \cdot 10^2 \leqslant \leqslant k_+^D \leqslant n \cdot 10^3$ ($\mu M^{-1}s^{-1}$), whilst for most kinases $n \leqslant k_+^D \leqslant n \cdot 10^2$ ($\mu M^{-1}s^{-1}$) (Schreiber *et al* 2009). If chosen randomly, $k_{+/-}$ are adjusted during numerical simulations in accordance with experimentally recorded data if and where available. The latter serve as model constraints and editing tools, being indispensable in making the particular choice of $k_{+/-}$ that lack direct *in vivo* or *in vitro* measurements (Neves 2011).

The results of numerical model simulations of the BBA are presented in the following paragraphs.

7.4 Simulation results

The assiduous reader must have realised that with so many enigmas and uncertainties apparent within the model it would be disingenuous to claim any quantitative prediction of the prototype's (the urinary bladder response) behaviour as accurate, even if it accidentally coincided with direct experimental measurements. Rather the results of numerical simulations are intended as proof of concept to demonstrate the feasibility of large-scale, multicomponent modelling of the BBA with regard to bladder wall growth and remodelling. Therefore, all results presented below are scaled to their attainable maximal absolute values as recorded during *in silico* experiments. The relative values are given in angle brackets. The reliability of the qualitative assessment of the biological processes under consideration, however, is assured by their associations with similar phenomena observed in other organs and systems in the body.

The results of numerical experiments to study the statics and dynamics of changes in shapes, force-stretch ratio distributions and electromechanical activity in the UB are discussed in detail elsewhere (Miftahof and Nam 2013, Miftahof and Akhmadeev 2019). Only a brief description of changes that initiate the cascade of the main electrochemical events along the BBA are presented here. The initial undeformed configuration of the bladder is determined by the amount of urine it contains at a given moment of time along with the intravesicular pressure, p. In simulations it has been assumed that at $t = 0$, the bioshell is loaded by $p = 1.0$ kPa. It is also assumed that the detrusor remains actively relaxed with no excitatory or inhibitory signals generated in the tissue. During filling, the body and apex of the bioshell undergo extensive deformations. Analysis of the total force distribution during this stage indicates that initially $p = 1.2$ kPa, with excessive tensions developing in a longitudinal direction in the body and apex of the organ. A high level of circumferential force is observed in the region of the flexure appearing along the posterior wall of the bladder. This extends bilaterally and upwards to the body of the organ. Low intensity longitudinal and circumferential stretch forces are seen in the fundus. With a gradual increase in p to 2.7 and 6.5 kPa, respectively, maximum T_l and T_c values are produced in the body and apex of the bioshell undergoing an even biaxial tension. A persistent zone of high degree T_c is recorded at the flexure. Throughout the process of filling, all regions of the organ experience biaxial stretching with $\lambda_{l,c} \gg 1.0$ (figures 7.2 and 7.3). This level of deformation is sufficient to trigger APs at the free nerve endings and dendrites.

The amplitudes and durations of the potentials recorded at the dendrites of the mechanoreceptors depend on the amplitudes and durations of the applied stimuli. Thus, a single stretch of the UB wall of intensity $\varepsilon = 0.1$ and duration 2.0 s is not enough to alter the receptor potential. The dendrites of the mechanoreceptors stay hyperpolarized. A gradual increase in the level of deformation leads to an increase in amplitude of the electrical response. For $\varepsilon = 0.4$, the deformation evokes a single

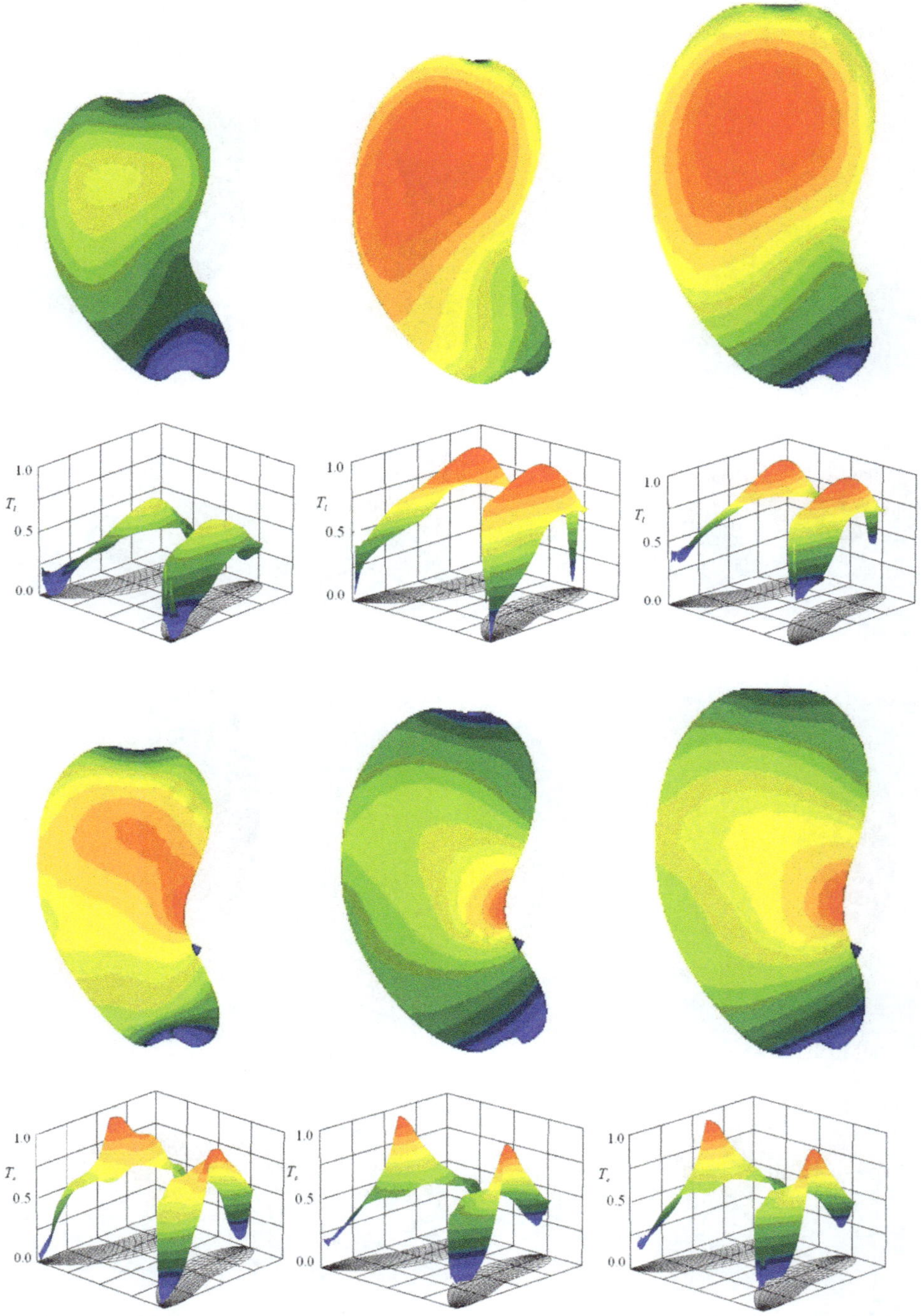

Figure 7.2. Changes in the longitudinal and circumferential T_l, T_c tension fields in the UB during filling. Reproduced from Miftahof and Nam (2013), with permission from Springer.

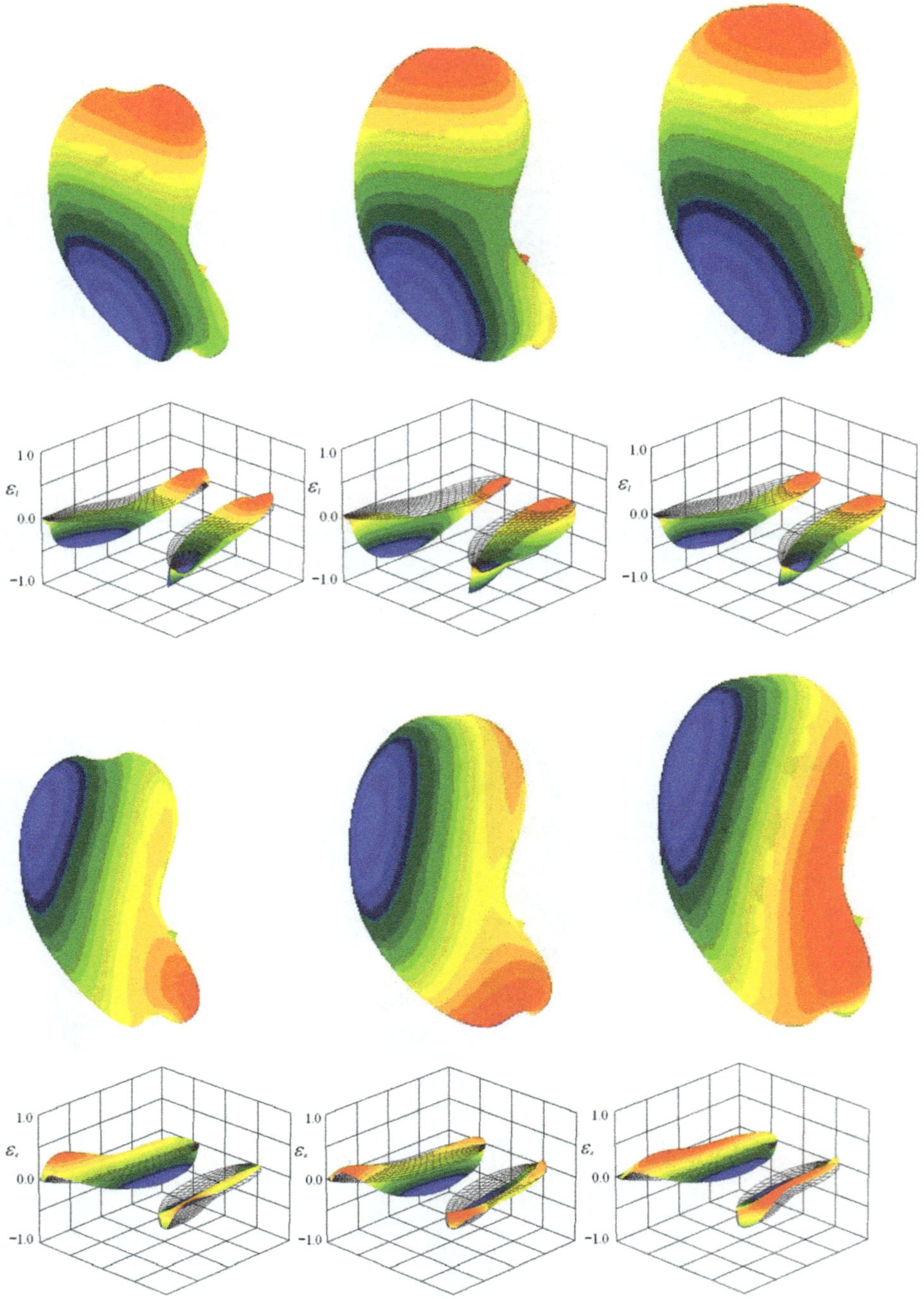

Figure 7.3. Changes in the axial and circumferential ε_l, ε_c tension fields in the UB during filling. Reproduced from Miftahof and Nam (2013), with permission from Springer.

receptor potential at the free nerve endings, $\langle V_{se}\rangle = 0.5$ and dendrites $\langle V_{sd}\rangle = 0.97$ of duration $\simeq 2.3$ ms (figure 7.4). A characteristic feature of the electrical pattern is the prolonged phase with a constant amplitude of approximately: $\langle V_{se}\rangle = 0.38$, $\langle V_{sd}\rangle = 0.7$, respectively.

The dendritic potential causes the development of an AP, $\max\langle V_{MS}\rangle = 1.0$, at the node of Ranvier of the hypogastric nerve (figure 7.5). This is conducted saltatory along the myelinated axon. On reaching the synaptic terminal at the axo-somatic synapse on the SDH neuron, it activates the voltage-dependent Ca^{2+} channel. This remains excited for a period of $\simeq 4$ ms allowing extracellular Ca^{2+} ions to enter the cell. The concentration of $\langle[Ca_i^{2+}]\rangle$ attains the maximum over the following 2 ms (figure 7.6(a)). A fraction of free ions passes rapidly to the synaptic vesicles whilst another is buffered intracellularly. On binding to Ca^{2+}-active centres on the vesicles, nearly 70% of Glu_v and $GABA_v$ are released to the synaptic cleft in free form (figures 7.6(b),(c)). A part of Glu_f and $GABA_f$ diffuse to the postsynaptic membrane, Glu_p and $GABA_p$, where they bind to their respective receptors and form biologically active complexes including: $Glu_p \cdot NMDA \cdot R \cdot Gly$, $Glu_p \cdot AMPA \cdot R$ and $GABA_p \cdot GABA_a \cdot R$ (figures 7.6(d),(e)). Other parts are either enzymatically degraded by GPT or transported by $GAT_{1/2/3}$ and $EAAT_3$ glutaminase to neurons for resynthesis (figure 7.6(f)). Activated ionotropic NMDA and AMPA receptors increase the permeability of voltage-dependent Ca^{2+} and Na^+ channels, respectively, whilst the $GABA_a$ receptor initiates a cascade of intracellular reactions leading to

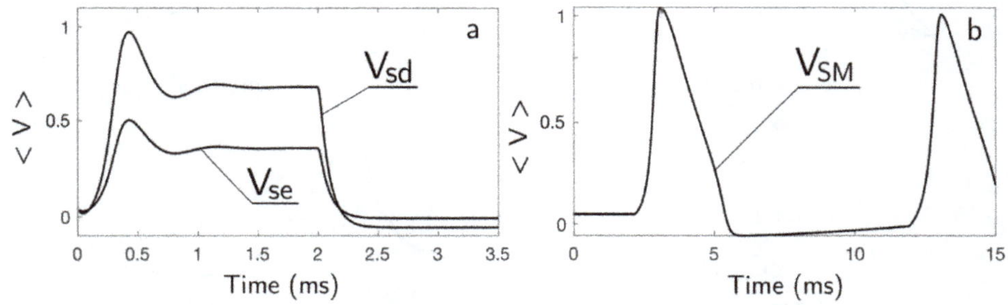

Figure 7.4. (a) Generation of free nerve ending and dendritic potentials at the mechano-receptor; (b) the subsequent production of action potentials at the node of Ranvier of the hypogastric sensory nerve fibre.

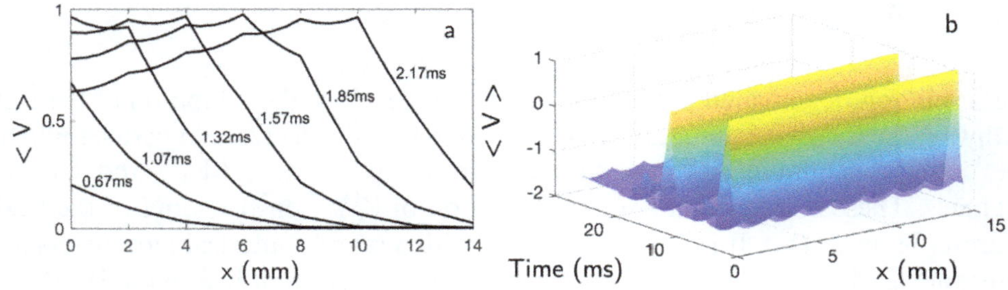

Figure 7.5. (a) The dynamics of saltatory conduction of the V_{MS} along the myelinated axon of the hypogastric nerve; (b) the time-space presentation of (a).

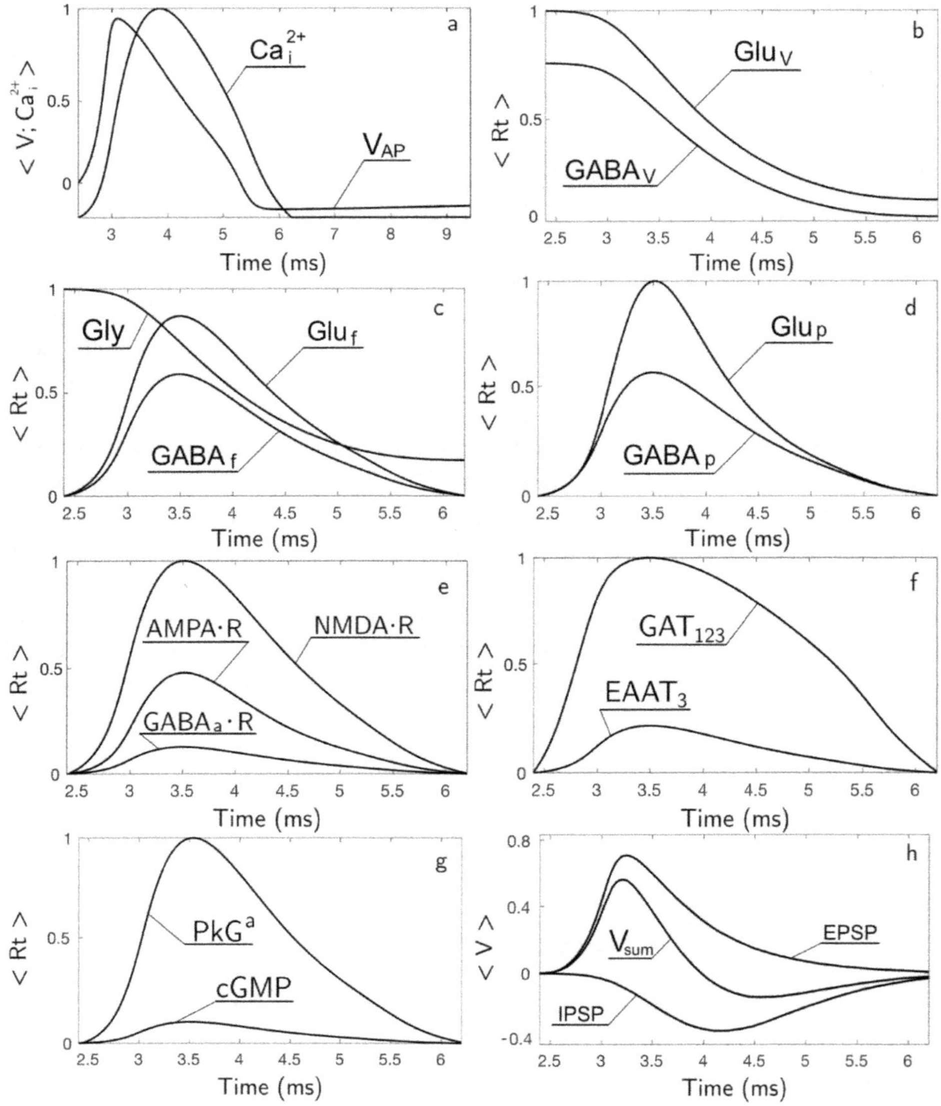

Figure 7.6. (a)–(h) Electrochemical signal transduction at the axo-somatic synapse on the SDH neuron. Here ⟨Rt⟩—respective reactant.

the formation of activated protein kinase G. The result is: the inhibition of the PLC pathway; the increase in reuptake of Ca_i^{2+} by the ER; the decrease in permeability of Ca^{2+}; the concomitant increase in the permeability of K^+, BK_{Ca} and Cl^- ion channels. These combined cause the generation of $V_{syn}^{(+/-)}$ on the soma of the SDH neuron (figure 7.6(h)). It is noteworthy that, with a set of numerical parameters and constants used in simulations, the EPSP reaches its maximum before the IPSP, and max $\langle V_{syn}^{+} \rangle$» max $\langle V_{syn}^{-} \rangle$.

The total postsynaptic potential, $\langle V_{syn}^{sum} \rangle = 0.6$, triggers the production of electrical discharges of maximal amplitude, $\langle V_{sdh} \rangle = 1$ of high frequency, $\nu \cong 75$ Hz, at the SDH neuron (figure 7.7(a)). Such dynamics secure the prolonged activation of transmembrane voltage-dependent channels, $\langle z \rangle$, and the influx of extracellular Ca^{2+} ions. The concentration of free calcium reaches its maximum level within 240 ms after excitation (figure 7.7(b)). The depolarisation of the neuron soma stimulates the generation and further propagation of the AP along the unmyelinated axon towards the PeVN neuron in the hypothalamus.

There is currently no electrophysiological data on PeVN neurons. Therefore, it has been assumed that the response to an excitatory stimulation from the SDH neuron resembles bursting type firing. Depolarisation of the membrane and the increase in influx of extracellular Ca^{2+} results in the release of free calcium ions from the ER. The rise in Ca_i^{2+} leads to exocytosis and extrasynaptic secretion of vesicular SS (figure 7.8(a)) with at least 65% of it released and nearly 80% reaching the ArcN-ME complex. Here it binds to $sst_{1/2}$ receptors. Due to the higher binding affinity of sst_1 receptors: $\langle SS_p \cdot sst_1 \rangle \gg \langle SS_p \cdot sst_2 \rangle$ (figure 7.8(b)). The concomitant activation of intracellular sGC and GPCR pathways result in a production of max $\langle sGC^a \rangle = 0.5$ and max $\langle G\gamma\beta \rangle = 1.0$ (figure 7.8(c)). These have multiple effects: the downregulation of adenylyl cyclase/cAMP/PkA signalling; the decrease in conductivity of transmembrane L- and T-type Ca^{2+} channels and Na^+/K^+ exchanger; the increase in permeability of K_{dr}, K_{ir} and BK_{Ca} ion channels, and the activation of PkG and the PLC pathway. The outcome is an inhibition in the secretion of GHRH and GH by ArcN-ME neurons and somatotrophs, respectively.

The subsequent electroneurohormonal excitation of the ArcN-ME neuron triggers the depolarisation of the cell membrane with the generation of high frequency APs. This is followed by a rise in $[Ca_i^{2+}]$ and GHRH release from LDCVs by extrasynaptic exocytosis (figure 7.9(a)). Somatostatin interferes with the dynamics by briefly affecting the process at $0.4 < \langle t \rangle < 0.6$. The neuron quickly recovers and max $\langle GHRH_b \rangle = 0.9$ enters the hypophysial portal blood system.

The binding of GHRH to its receptors located on the somatotroph soma leads to the activation of the G_s-protein-coupled system and the AC/cAMP/PkA pathway (figures 7.8(b),(c)). The rapid formation of $\langle G\gamma\beta \rangle = 1.0$ and the production of $\langle PkA^a \rangle = 0.35$ result in the increase in conductivity of transmembrane Na^+, VGCa,

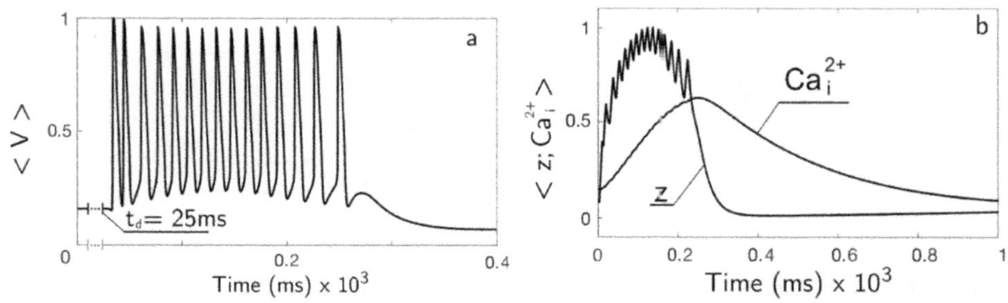

Figure 7.7. (a) The response of the SDH neuron to the total postsynaptic potential; (b) the dynamics of activation of transmembrane Ca^{2+} ion channels (z) and the rise of intracellular calcium.

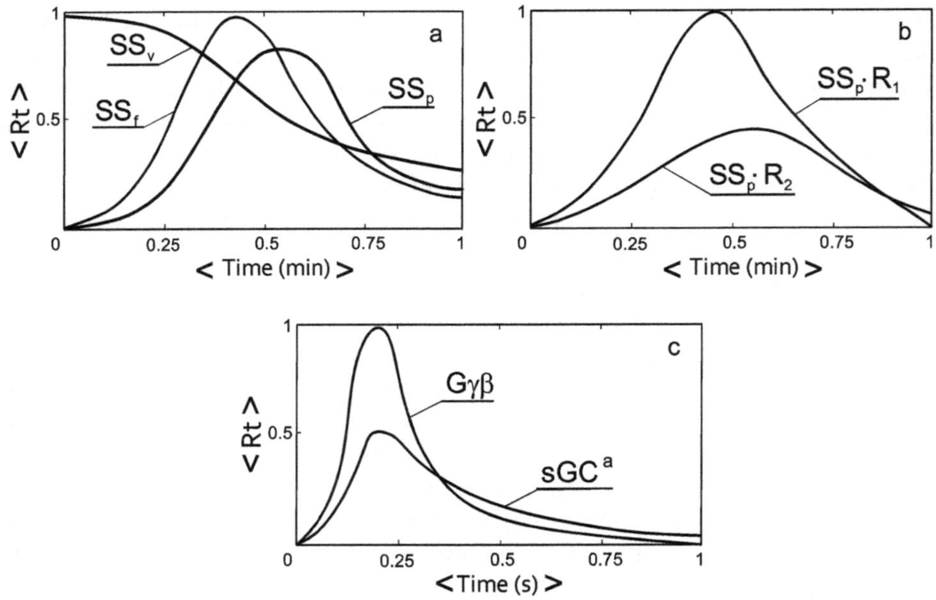

Figure 7.8. (a)–(c) The response of the PeV neuron with somatostatin release, the activation of relevant receptors on the somas of neurons of the ArcN-ME complex, and the production of reactive substrates. Here $\langle SS_p \cdot R_1 \rangle$, $\langle SS_p \cdot R_2 \rangle$—SS-sst$_1$ and SS-sst$_2$ complexes.

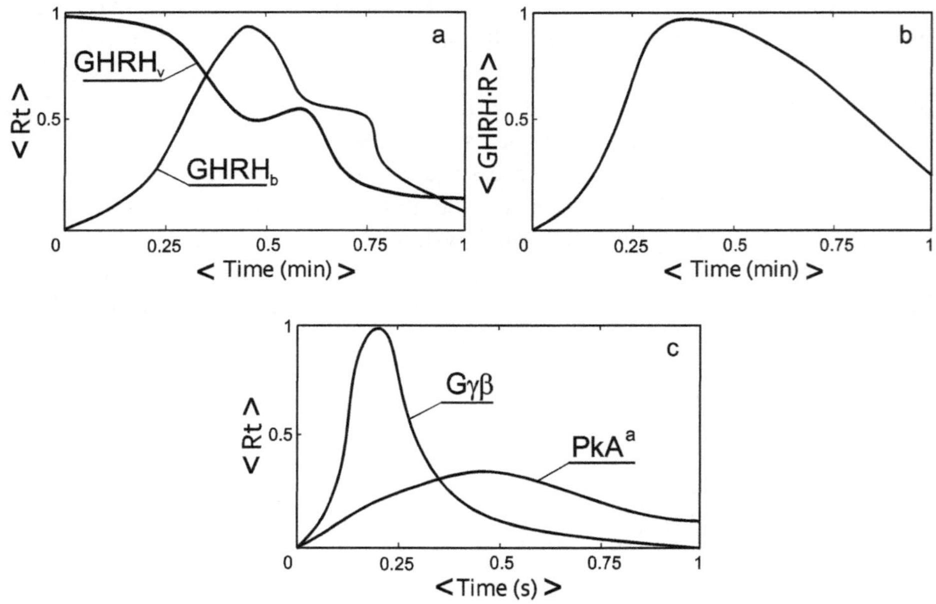

Figure 7.9. (a)–(c) The ArcN-ME complex response with GHRH release, activation of GHRH receptors on the somatotroph, and the production of reactive substrates.

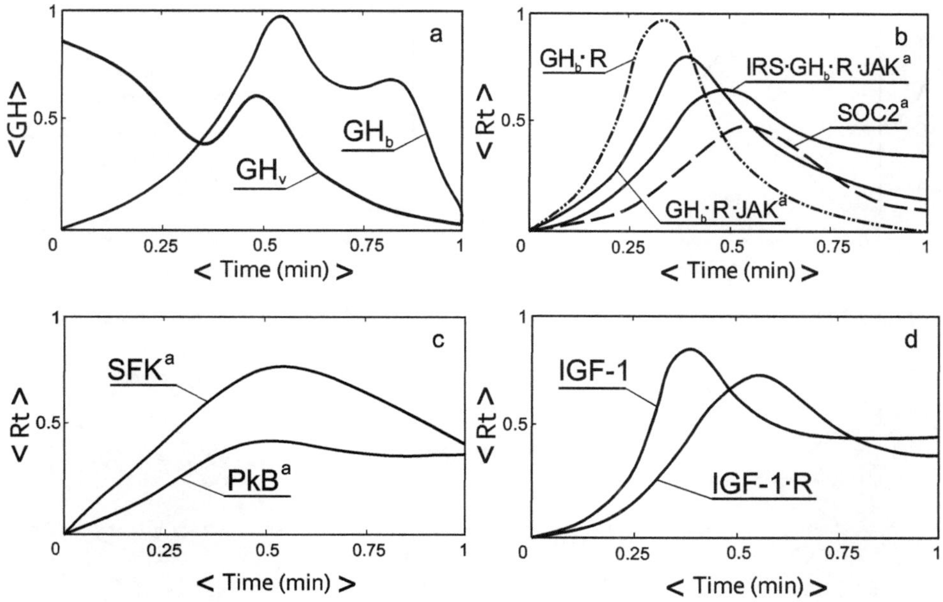

Figure 7.10. (a)–(d) The dynamics of the release of GH by somatotrophs of the anterior pituitary gland and the production of IGF-1 by hepatocytes.

L- and T-type Ca^{2+} channels, and upregulation of GH secretion by the somatotroph.

Depolarisation of the somatotroph with the generation of APs, $\langle V_{st} \rangle = 0.9$, of frequency 0.5 Hz, induces the extracellular influx of Ca^{2+} ions with a concomitant production of DAG and IP_3. The subsequent activation of ryanodine receptors on the endoplasmic reticulum stimulates the release of Ca^{2+} ions from the ER store into the cytosol in a pulsatile manner. The rise in $[Ca_i^{2+}]$, the effect of activated GHRH · R, trigger the exocytosis of GH from LDCVs into the blood circulation (figure 7.10(a)). The major part of the hormone combines quickly with GH-binding serum protein, $\langle GH_b \cdot GHBP \rangle = 0.95$, whilst the remainder binds to its receptors, $\langle GH_b \cdot R \rangle = 0.95$, on the target tissues—hepatocytes and the UB wall. Activation of tyrosine kinase-dependent regulatory pathways follows with the formation of biologically active complexes such as $\langle GH_b \cdot R \cdot JAK^a \rangle = 0.8$, $\langle IRS \cdot GH_b \cdot R \cdot JAK^a \rangle = 0.6$, $\langle PkB^a \rangle = 0.4$, $\langle SOC2^a \rangle = 0.45$, and $\langle SFK^a \rangle = 0.75$ (figures 7.10(b),(c)). These instigate the production of IGF-1 which, upon binding to their receptors, $\langle IGF{-}1 \cdot R \rangle = 0.7$, (figure 7.10(d)) together with activated $GH_b \cdot R$, lead to gene expression. The process involves ribosomal kinase S6 activation, $\langle S6^a \rangle = 0.9$, with the ultimate induction of collagenesis and myogenesis (figure 7.11). During the growth of the urinary bladder wall, more smooth muscle than connective tissue is produced, $\langle SM \rangle = 0.6$, $\langle Col \rangle = 0.35$. This can be expected to stiffen the wall rendering it less compliant.

Clinical observations indicate that the UB accommodates and retains large quantities of urine in cases of urinary outflow obstruction. The results of simulations

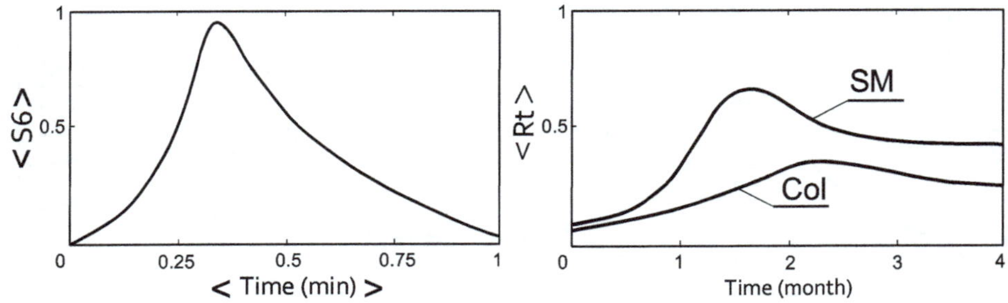

Figure 7.11. Gene activation and induction of collagenesis and myogenesis in bladder wall tissue.

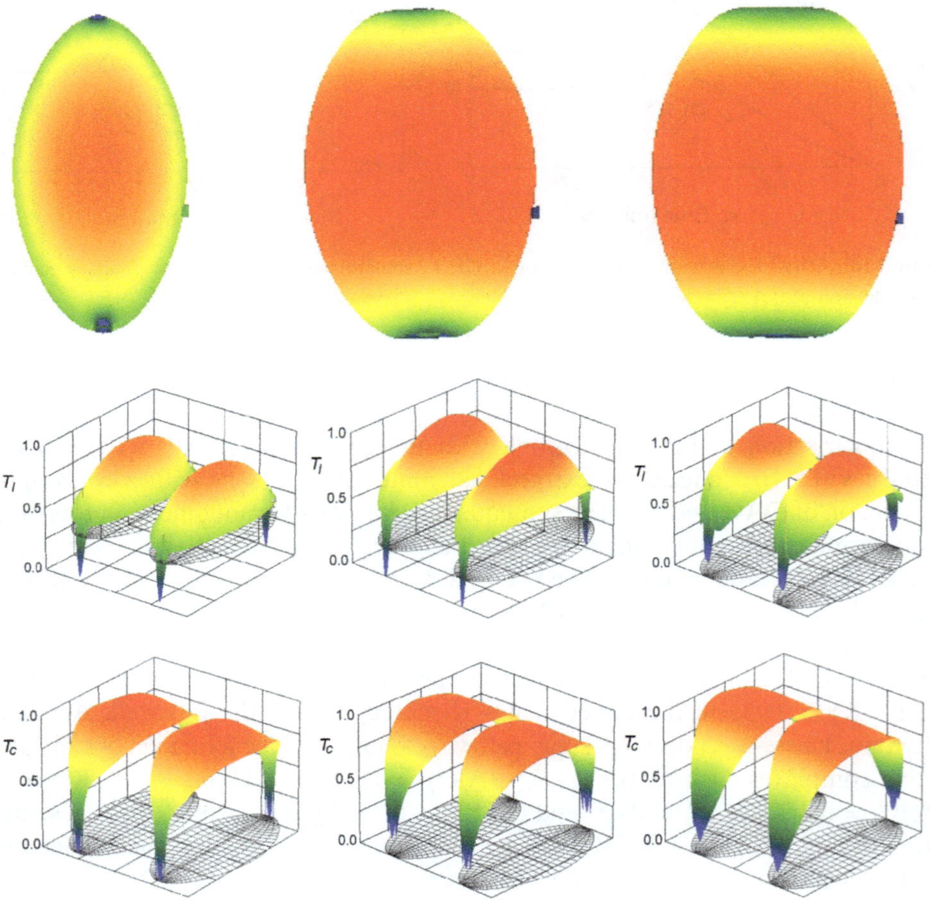

Figure 7.12. Changes in the longitudinal and circumferential T_l, T_c tension fields in the UB during filling after growth and remodelling.

conducted to analyse the dynamics of change in shapes and plain force distribution in the UB after G&R are given below. The initial configuration of the bladder differs from that shown in figure 7.2 and is a result of tissue G&R. During the numerical experiment, the mechanical characteristics of the bladder wall have been adjusted to reflect G&R whilst the loading conditions remain constant. It is also assumed that the detrusor remains relaxed with no excitatory or inhibitory signals generated in the tissue. During filling and storage phases, the bioshell undergoes isomorphic deformations, i.e. attaining shapes that do not differ in their appearance (figure 7.12). The total force distribution during these stages indicates excessive tension developing in both longitudinal and circumferential directions in the body of the organ with lower intensity stretch forces seen in the apex and fundus. A gradual increase in p, leads to an increase in the maximum values of T_l and T_c. All regions of the organ experience biaxial stretching.

The reader should also note that the results of these numerical simulations describe the response of the UBA to one single round of stimulation only, i.e. the distension of the bladder wall leading to the growth reaction of the tissue. More interesting phenomena could be expected related to the multiple presence and conjoint action of electrical events, neurotransmitters and hormones found within the BBA (for details see figure 7.1(b)) that follow continuous repetitive excitation during the course of outflow obstruction. This has not been investigated here and is left to the diligent researcher to develop as a future project.

References

Amaro R E, Ieong P, Huber G, Dommer A, Steven A C, Bush R M, Durrant J D and Votapka L W 2018 A computational assay that explores the hemagglutinin/ neuraminidase functional balance reveals the neuraminidase secondary site as a novel anti-influenza target *ACS Cent. Sci.* **4** 1570–7

Babtie A C and Stumpf M P H 2017 How to deal with parameters for whole-cell modelling *J. R. Soc. Interface.* **14** 20170237

Cox D 2006 *Principles of Statistical Inference* (Cambridge: Cambridge University Press)

ElSawy K M, Lane P, Verma C S and Caves L S D 2016 Recognition dynamics of p53 and MDM2: implications for peptide design *J. Phys. Chem.* B **120** 320–8

Ermak D and McCammon J 1978 Brownian dynamics with hydrodynamic interactions *J. Chem. Phys.* **69** 1352–60

Fröhlich F, Theis F J and Hasenauer J 2014 Uncertainty analysis for nonidentifiable dynamical systems: profile likelihoods, bootstrapping and more *Computational Methods in Systems Biology (Lecture Notes in Computer Science* vol 8859) ed P Mednes *et al* pp 61–72

Gabdoulline R R and Wade R C 2001 Protein–protein association: investigation of factors influencing association rates by Brownian dynamics simulations *J. Mol. Biol.* **306** 1139–55

Gabdoulline R R and Wade R C 2002 Biomolecular diffusion association *Curr. Opin. Struct. Biol.* **12** 204–13

Geyer T and Winter U 2009 An O(N_2) approximation for hydrodynamic interactions in Brownian dynamics simulations *J. Chem. Phys.* **130** 114905

Haddadian E J and Gross E L 2006 Brownian dynamics study of the interactions of the luminal domains of the cytochrome b6f complex with plastocyanin and cytochrome c6: the effects of the reiske fes protein in the interactions *Biophys. J.* **91** 2589–600

Hall M P *et al* 2012 Engineered luciferase reporter from a deep sea shrimp utilizing a novel imidazopyrazinone substrate *ACS Chem. Biol.* **7** 1848–57

Hirakis S P, Malmstrom R D and Amaro R E 2017 Molecular simulations reveal an unresolved conformation of the type IA protein kinase A regulatory subunit and suggest its role in the cAMP regulatory mechanism *Biochem. J.* **56** 3885–8

Huber G A and McCammon J A 2019 Brownian dynamics simulations of biological molecules *Trends Chem.* **1** P727–38

Jagiedella N, Rickert D, Theis F J and Hasenauer J 2017 Parallelization and high-performance computing enables automated statistical inference of multiscale models *Cell. Sys.* **4** 194–206

Kilpatrick L E *et al* 2017 Real-time analysis of the binding of fluorescent $VEGF_{165}a$ to VEGF2 in living cells: effect of receptor tyrosine kinase inhibitors and fate of internalized agonist-receptor complexes *Biochem. Pharmacol.* **136** 62–75

Liepe J, Kirk P D W, Filippi S, Toni T, Batnes C P and Stumpf M P H 2014 A framework for parameter estimation and model selection from experimental data in systems biology using approximate Bayesian computation *Nat. Protoc.* **9** 439–56

McGuffee S and Elcock A 2010 Diffusion, crowding & protein stability in a dynamic molecular model of the bacterial cytoplasm *PLoS Comput. Biol.* **6** e1000694

Moles C G, Mendes P and Banga J R 2003 Parameter estimation in biochemical pathways: a comparison of global optimization methods *Genome Res.* **13** 2467–74

Motulsky H J and Mahan L C 1984 The kinetics of competitive radioligand binding predicted by the law of mass action *Mol. Pharmac.* **25** 1–9

Miftahof R N and Nam H G 2013 *Biomechanics of the Human Urinary Bladder* (Heidelberg: Springer)

Miftahof R N and Akhmadeev N R 2019 *Soft Biological Shells in Bioengineering* (Bristol: IOP Publishing)

Neves S 2011 Obtaining and estimating kinetic parameters from the literature *Sci. Signal* **4** tr8

Noé F, Schütte C, Vanden-Eijnden E, Reich L and Weikl T R 2009 Constructing the full ensemble of folding pathways from short term equilibrium simulations *Proc. Natl. Acad. Sci. USA* **106** 19011–6

Northrup S H, Allison S A and McCammon A J 1984 Brownian dynamics simulation of diffusion influenced bimolecular reactions *J. Chem. Phys.* **80** 1517–24

Ozturk M A, Cojocaru V and Wade R C 2018 Dependence of chromatosome structure on linker histone sequence and posttranslational modification *Biophys. J.* **114** 2363–75

Pachov G V, Gabdoulline R R and Wade R C 2011 On the structure and dynamics of the complex of the nucleosome and the linker histone *Nucleic Acids Res.* **39** 5255–63

Pang X D and Zhou H X 2016 Mechanism and rate constants of the Cdc42 GTPase binding with intrinsically disordered effectors *Proteins-Struct. Funct. Bioinform.* **84** 674–85

Peach C J, Kilpatrick L E, Friedman-Ohana R, Zimmerman K, Robers M B, Wood K V, Woolard J and Hill S J 2018 Real-time ligand binding of fluorescent VEGF-A isoforms that discriminate between VEGFR2 and NRP1 in living cells *Cell. Chem. Biol.* **25** 1208–18

Raue A, Kreutz C, Maiwald T, Bachmann J, Shilling M, Klugmüller U and Timmer J 2009 Structural and practical identifiability analysis of partially observed dynamical models by exploiting the profile likelihood *Bioinform. J.* **25** 1923–29

Romanowska J, Kokh D B and Wade R C 2015 When the label matters: adsorption of labeled and unlabeled proteins on charged surfaces *Nano Lett.* **15** 7508–13

Schiele F, Ayaz P and Fernández-Montalávan A 2015 A universal homogenous assay for high-throughput determination of binding kinetics *Ann. Biochem.* **468** 42–9

Schreiber G, Haran G and Zhou H-X 2009 Fundamental aspects of protein–protein association kinetics *Chem. Rev.* **109** 839–60

Smoluchowski M V 1917 Versuch einer mathematischen theorie der koagulationskinetik kolloider losungen *Z. Phys. Chem.* **XCII** 129–68

Šolc K and Stockmayer W H 1973 Kinetics of diffusion-controlled reaction between chemically asymmetric molecules. II. Approximate steady-state solution *Int. J. Chem. Kinet.* **5** 733–52

Sun J, Garibaldi J M and Hodgman C 2012 Parameter estimation using meta-heuristics in systems biology: a comprehensive review *IEEE/ACM Trans. Comput. Biol. Bioinform.* **9** 185–202

Sykes D A, Stoddart L A, Kilpatrick L E and Hill S J 2019 Binding kinetics of ligands acting at GPCRs *Mol. Cell. Endocrin.* **485** 9–19

Xia L, de Vries H, Ijzerman A P and Heitman L H 2016 Scintillation proximity assay (SPA) as a new approach to determine a ligand's kinetic profile. A case in point for the adenosine A1 receptor *Purinergic Sign.* **12** 115–26

Zamora-Sillero E, Hafner M, Ibig A, Stelling J and Wagner A 2011 Efficient characterization of high-dimensional parameter spaces for systems biology *BMC Syst. Biol.* **5** 142

Zhou H X 1993 Brownian dynamics study of the influences of electrostatic interaction and diffusion on protein–protein association kinetics *Biophys. J.* **64** 1711–26

Zhou H X 1997 Enhancement of protein–protein association rate by interaction potential: accuracy of prediction based on local Boltzmann factor *Biophys. J.* **73** 2441–5

IOP Publishing

The Brain–Bladder Axis in Tissue Growth and Remodelling

Roustem N Miftahof and Christian J Cyron

Chapter 8

What is to follow?

There is no way mathematical modelling could solve major biological problems on its own. On the other hand, it is highly unlikely that even a reasonably complete understanding could come solely from experiment.

— J D Murray

8.1 Making a model reliable

Systems computational biology is genuinely interdisciplinary in that it covers a broad spectrum of topics that extend from: gene regulation to tissue growth and remodelling; molecular pharmacology to drug discovery; collagen synthesis to wound healing; the pathophysiology of diseases, to mention just a few. The basic philosophy behind it is to truthfully reconstruct a sequence of events using meaningful and experimentally measurable quantities which can be interrogated mathematically, either analytically or numerically, with the results reinterpreted biologically (Cochran 1961, Murray 2012). Hence it is expected that a computational biologist has a thorough knowledge of biosciences and mathematics, as well as the aptitude to efficiently integrate ideas and results between the two.

Mathematical models in biology rely on assumptions contingent on the laws and principles by which living systems operate—the laws of fundamental physics—and the researcher has to assimilate and apply the latter to describe objectively the processes to hand. All assumptions ordinarily ensconce 'convenient' simplifications and it is the responsibility of a sagacious modeller to make and choose ones that carry sufficient acumen. A model whose assumptions correspond approximately to reality will make predictions that are approximately valid (Huppert and Kartiel 2013). Hence even simplified models, as long as their assumptions reflect some central aspects of reality, can provide worthy predictions. The validity and relevance of the approach has been demonstrated in early models of cell metabolism, nerve pulse conduction, gene expression, etc, that correctly replicate and facilitate an

doi:10.1088/978-0-7503-3567-6ch8

understanding of the fundamental aspects of molecular and cell physiology within a coherent mathematical framework (Zavala *et al* 2019).

To capture the essential behavioural features of a complex system such as the BBA in a single computationally tractable framework, a model must integrate several morphostructural elements including the urinary bladder, the dorsal root ganglia, the periventricular nuclei, the arcuate-median eminence complex, the adenohypophysis and hepatocytes along with the forward/feedback interactions amongst them and extensive details of the intracellular signalling pathways. Such a model is multicompartmental, multiscale, large in terms of parameters and hybrid in nature (Walpole *et al* 2013, Le Novère 2015, Hasenauer *et al* 2015, Babtie and Stumpf 2017). It crystallises a concept of the bladder–brain interaction in a set of mathematical equations which can only be analysed numerically (computationally).

Caution needs to be exercised, however, when using complex models. Even if individual compartments are accurately characterised and carefully parameterised, the scale-integration of subsystems in a compound model may introduce uncertainties and errors. The former, for example, might be related to cellular heterogeneity and connectivity, whilst the latter can be caused by averaging specie-dependent parameters, the variability of input and receptive fields amongst others. These matters of biological realism are difficult to treat purely mathematically through the simple addition of either weight factors, penalty functions or artificial rate constants. Bridging the time and spatial scales of subsystems, which may vary from millisecond to hours/days and from nanometres to centimetres/metres, while ensuring the consistency of the compound model is the only way to: reduce uncertainties; assess and alleviate the propagation of errors across the model; find suitable mathematical approximations, etc. If ignored, these will render simulation results intractable to correct interpretation (Gerstner *et al* 2012, Villaverde and Banga 2013, Liepe *et al* 2013).

Mathematics is highly versatile, adaptable and pervasive throughout practically every field of modern science. However, it is not always applied either tactfully or skilfully (May 2004). Deciding on the uniqueness and robustness of a model is a formidable challenge. Currently this is achieved by comparing simulation results with existing experimental data, i.e. an experimentally biased evaluation. An unbiased approach is to test the model against other models that include more or less elaborate working assumptions and descriptions of processes. If the results ensure their independence of assumptions, parameterisation and the numerical algorithm, then the robustness and fidelity of the model in question is proven. True judgement, however, should be made based on an assessment that involves both strategies. Unfortunately, since most of the models in biology are created to reproduce and fit individual experiments, the former is preferred to the latter which are often unavailable and 'wittingly concealed' from the public, i.e. the structure and design of software along with the execution of the numerical algorithms used are not revealed by their creators. Besides, modern day computational biology relies heavily on commercial simulation software usually supported by user attractive graphic tools. In the hands of a fervent researcher driven by a desire to obtain a result at any cost, deceptive results can easily follow.

Another caveat is related to the parameterisation of the model, i.e. parameter identification, initialisation, estimation and sensitivity. If the model is to include only experimentally verified parameters, there is a risk of biasing it according to particular experimental limitations. Sadly, many systems biology models contain parameters that are unknown and thus have to be deduced from *in silico* simulations, that is to say, provided empirical observation results are known, the inverse problem to find parameter sets that produce a good fit for the model output with the experimental data has to be solved. In general, these problems are ill-posed, subjected to data noise and require initial parameter values which are sparse. Thus in the proposed BBA model, structural identifiablity, e.g. *a priori* information on the cellular, synaptic and blood concentrations of neuro-transmitters and hormones, the receptor distribution and the degree of crosstalk among intracellular signalling pathways, at any time, is incomplete and limited. As such, it relies mainly on practical identifiability—experimental data.

As the biological processes are strongly correlated, so are the parameters that quantify them. With a large model, a global rather than a local sensitivity analysis of the multidimensional parameter space helps to calculate the propagation of uncertainties and errors, provide a comprehensive assessment of the degree of dependence between parameters, and select sets of parameters that exert the maximum effect on model responses. It is likely that a sensitivity analysis can provide a number of datasets that replicate the observed pattern of behaviour. The selection of one that best reflects reality needs to be based on comparison with judiciously designed, discriminatory experiments. At the time of writing, justification, assertion and thorough test-analysis of the validity, limitations and feasibility of the BBA model could not be provided by the authors.

Finally, despite the fact that mathematical abstraction expresses intuition within a unified coherent framework (models), all models are biased towards our current interpretation of reality, which depends heavily on experimental observations. Consequently, modelling can be viewed as an iterative process synergistic with experimentation, one driving the other. Models as such must be continuously refined and improved to become more accurate and more reliable in accordance with recent advancements in technology and the continuing change in our perception of life.

8.2 Model expansions in biomedicine

Good model building is a subjective exercise. The recent explosion in bioinformatics enabling inference and reliable estimations to provide data previously missing has dispelled cynicism and outdated perceptions concerning the usefulness and the ‘thorny questions of identifiability’ of complex biological systems models (Keener and Sneyd 2009). However, since ‘models are not reality and they are not meant to represent all aspects of life faithfully’, this leaves the notion of constructing a comprehensive, detailed model superfluous (Black 1993, Gunawardena 2014, Babtie and Stumpf 2017).

The focus of this book has been to establish and investigate the mechanobiological aspect of growth and remodelling of the wall of the urinary bladder (UB) as a result of urinary outflow obstruction. It has implicitly assumed the presence of urine in the UB which has been introduced into the model by a single parameter—intravesicular pressure p. However, the explicit inclusion of urine becomes essential in problems related to the quantification, assessment, diagnosis and establishment of the efficacious treatment of patients with abnormal lower urinary tract dynamics. This notion is supported by experimental evidence that a dense nexus of suburothelial and urothelial nerves located in the bladder is critical to the sensory function (Birder *et al* 2001, Wiseman *et al* 2002). Traditionally viewed merely as a passive barrier, it has become obvious that the urothelium actively responds to exogenous/endogenous mechanical and chemical stimuli, transmitting both across bladder wall (de Groat 2004). Hence, one of the challenges that still remains unsolved in the biomechanics of the UB is the mathematical formulation and solution to the problem of the urine-bladder–brain axis (Miftahof and Nam 2013). Although several attempts have been made to model urine-bladder wall interactions, all have been limited to an analysis of the patterns of urinary velocities produced as a result of prescribed wall movements (Tziannaros and Smith 2012).

The proposed BBA model extension could be viewed as follows.

A medium where the stress vector $\overline{p}_n$ at any point $\forall\ M \in S$ coincides with the direction of the normal $\overline{n}$ to the surface S is called an ideal fluid. In this case the tensor surface is a sphere and the principal components of the stress tensor are equal to $p_1 = p_2 = p_3 = -p$, where p is the (mechanical) pressure.

A fluid for which the stress tensor is given by

$$p^{ij} = -pg^{ij} + \tau_u^{ij}, \qquad (i, j = 1,2,3) \tag{8.1}$$

where

$$p = p(\rho_u, T, \chi_1, \dots, \chi_n),$$
$$\tau_u^{ij} = f^{ij}(\dot{\varepsilon}_{\alpha\beta}, g^{ij}, \chi_1, \dots, \chi_n),$$

is called a viscous fluid. Here ρ_{u} is the density of fluid, T is temperature, $\tau(\tau_u^{ij})$ is the shear stress exerted by the fluid, $\dot{\varepsilon}_{\alpha\beta}$ are the covariant components of the stress and of the rate of deformation tensors, respectively, g^{ij} are contravariant components of the metric tensor, and $\chi_1, \dots, \chi_n$Z are physicochemical parameters, e.g. the fraction of solutes in urine.

Assume the fluid satisfies the condition: $p^{ij} = 0 \Rightarrow \dot{\varepsilon}_{\alpha\beta} = 0$. Then expanding f^{ij} in a Taylor series with respect to $\dot{\varepsilon}_{\alpha\beta}$ and neglecting non-linear terms, for τ_u^{ij} we obtain

$$\tau_u^{ij} = A^{ij\alpha\beta}\dot{\varepsilon}_{\alpha\beta}, \tag{8.2}$$

which is the general form of Navier–Stokes's law for viscous Newtonian fluids. The total number of components of the tensor of intrinsic mechanical properties $A^{ij\alpha\beta}$ is 81. However, taking into consideration the symmetry of the stress and the rate of deformation tensors, $p^{ij} = p^{ji}$, $\varepsilon_{\alpha\beta} = \varepsilon_{\beta\alpha}$, the number of $A^{ij\alpha\beta}$ can be reduced to 36. In the case of isotropic fluid, only 9 non-zero components remain: $A^{1111} = A^{2222} = A^{3333}$, $A^{1122} = A^{1133} = A^{2233}$, $(A^{ii\alpha\alpha} = A^{\alpha\alpha ii}$, $A^{ijij} \neq 0$, $i \neq j)$.

Introducing parameters μ_u, λ_u, such that $A^{iiii} = 2\mu_u + \lambda_u$, $A^{ii\alpha\alpha} = \lambda_u$, expression (8.2) can be written in the principal axes of the stress and the rate of deformation tensors, $\dot{\varepsilon}_i$ and $\tau_{(u)i}$, as

$$\tau_{(u)i} = \lambda_u \sum_{j=1}^{3} \dot{\varepsilon}_j + 2\mu_u \dot{\varepsilon}_i, \qquad i = 1,2,3. , \tag{8.3}$$

Multiplying equation (8.3) by $(dx^i)^2$ and, after simple rearrangements, Navier–Stokes' law in the general curvilinear coordinate system takes the form (Sedov 1976)

$$\tau_{(u)}^{ij} = \lambda_u I_1(\dot{\varepsilon}) g_{ij} + 2\mu_u \dot{\varepsilon}_{ij}, \tag{8.4}$$

and for the contravariant components

$$\tau_{(u)}^{ij} = \lambda_u I_1(\dot{\varepsilon}) g^{ij} + 2\mu_u g^{i\alpha} g^{j\beta} \dot{\varepsilon}_{\alpha\beta}. \tag{8.5}$$

Substituting (8.5) into (8.1) for an isotropic viscous fluid in the general curvilinear coordinate system, we get

$$p^{ij} = -pg^{ij} + \lambda_u g^{ij} div \bar{v}_u + 2\mu_u \dot{\varepsilon}^{ij}, \tag{8.6}$$

where $I_1(\dot{\varepsilon}) := div\bar{v}_u$ and $\bar{v}_u(u_u, v_u, w_u)$ is the velocity vector.

The fluid is considered to be incompressible if an arbitrary volume of it remains constant throughout the space and time. From the law of the conservation of mass, it follows

$$\frac{d\rho_u}{dt} = \frac{\partial \rho_u}{\partial t} + \bar{v}_u \cdot \nabla \rho_u = \frac{\partial \rho_u}{\partial t} + \nabla \cdot (\rho_u \bar{v}_u) = 0, \tag{8.7}$$

where $d\rho_u dt$ is the material derivative and the meaning of other parameters is as described above. In the case of homogenous (ρ_u = const), incompressible fluid, ρ_u becomes a single-valued parameter and equation (8.7) takes the simplest form

$$\nabla \cdot \bar{v}_u = 0, \tag{8.8}$$

which is the statement of the conservation of volume.

To derive the system of Navier–Stokes equations that describes the dynamics of viscous incompressible Newtonian fluid, we proceed with the law of motion. In the general curvilinear coordinate system, it is given by

$$\rho_u \bar{a}^j = \rho_u \bar{f}^j + \nabla_i p^{ji}, \tag{8.9}$$

where $\bar{a}^j$ is acceleration

$$\bar{a}^j = \frac{\partial v^j}{\partial t} + v^i \nabla_i v^j = \frac{\partial v^j}{\partial t} + v^i \left(\frac{\partial v^j}{\partial x^i} + v^s \Gamma_{si}^j \right). \qquad (.\ 8.10)$$

Here $\bar{f}^j$ is the density of the mass force, ∇_i is the sign of the covariant derivative, and Γ_{si}^j are the Christoffel symbols of the second kind. Covariant differentiation of (8.6) with respect to p^{ij}, yields

$$\nabla_i p^{ij} = -\nabla^i p + (\lambda_u + \mu_u)\nabla^i(\nabla \cdot \bar{v}_u) + \mu_u \Delta v^i,$$

where ∇^i, ∇ are the contravariant derivative and the Laplace operator, respectively. In the above, we have assumed that λ_u and μ_u are constants. Additionally, if the fluid is incompressible then

$$\nabla_i p^{ij} = -\nabla^i p + \mu_u \Delta v^i. \tag{8.11}$$

Substituting (8.11) in (8.9), we obtain

$$\rho_u \bar{a}^j = \rho_u \bar{f}^j - \nabla^i p + \mu_u \Delta v^i. \tag{8.12}$$

Equations (8.12), (8.8) constitute the governing system of the equations of motion of viscous incompressible Newtonian fluid in the general curvilinear coordinate system with constant dynamic viscosity μ_u. The formulation of pertinent initial and boundary conditions closes the problem.

In cylindrical coordinates $\{r, \varphi, z\}$, equations (8.12) and (8.8) take the form (Sedov 1976)

$$\begin{aligned}
\rho_u \bar{a}_r &= \bar{f}_r - \frac{\partial p}{\partial r} + \mu_u\left(\Delta u_r - \frac{u_r}{r^2} - \frac{2}{r^2}\frac{\partial v_\varphi}{\partial \varphi}\right),\\
\rho_u \bar{a}_\varphi &= \bar{f}_\varphi - \frac{1}{r}\frac{\partial p}{\partial r} + \mu_u\left(\Delta u_\varphi - \frac{u_\varphi}{r^2} + \frac{2}{r^2}\frac{\partial u_r}{\partial \varphi}\right),\\
\rho_u \bar{a}_z &= \bar{f}_z - \frac{\partial p}{\partial z} + \mu_u \Delta u_z
\end{aligned} \tag{8.13}$$

and

$$r\frac{\partial \rho_u}{\partial t} + \frac{\partial \rho_u u_r r}{\partial r} + \frac{\partial \rho_u v_\varphi}{\partial \varphi} + r\frac{\partial \rho_u w_z}{\partial z} = 0. \tag{8.14}$$

Here, the Laplacian is given by

$$\Delta = \frac{\partial^2}{\partial r^2} + \frac{1}{r}\frac{\partial}{\partial r} + \frac{1}{r^2}\frac{\partial^2}{\partial \varphi^2} + \frac{\partial^2}{\partial z^2}$$

and the components of the acceleration vector are

$$\begin{aligned}
\bar{a}_r &= \frac{\partial u_r}{\partial t} + u_r\frac{\partial u_r}{\partial r} + \frac{v_\varphi}{r}\frac{\partial u_r}{\partial \varphi} + w_z\frac{\partial u_r}{\partial z} - \frac{v_\varphi^2}{r},\\
\bar{a}_\varphi &= \frac{\partial v_\varphi}{\partial t} + u_r\frac{\partial v_\varphi}{\partial r} + \frac{v_\varphi}{r}\frac{\partial v_\varphi}{\partial \varphi} + w_z\frac{\partial v_\varphi}{\partial z} + \frac{u_r v_\varphi}{r},\\
\bar{a}_z &= \frac{\partial w_z}{\partial t} + u_r\frac{\partial w_z}{\partial r} + \frac{v_\varphi}{r}\frac{\partial w_z}{\partial \varphi} + w_z\frac{\partial w_z}{\partial z}, \quad \bar{v}_u(u_r, v_\varphi, w_z).
\end{aligned}$$

Urine is approximately 95% water with the remaining 5% dissolved non-organic and organic solutes. These include Na^+, K^+ ,Ca^{2+} ,Mg^{2+} ,Cl^-, ammonium, phosphate and sulphate ions, along with urea, creatinine and uric acid. Their concentration depends on metabolic state, diet and administered medications. Normal urine density is $\rho_u \simeq$ 1.003–1.035 (g cm^{-3}), while that of water is, $\rho_w \simeq 0.998$ (g cm^{-3}). Any deviations are associated with urinary disorders, e.g. $\rho_u < 1.003$ g cm^{-3} is observed in overhydration or anti-diuretic hormone deficiency whilst $\rho_u > 1.035$ g cm^{-3} is commonly seen with albuminuria, glucosuria, pyuria, etc. These variations are linearly related to the solute weight fraction, [...], (Putnam 1971)

$$\rho_u = 0.04775[...] + 0.99325. \tag{8.15}$$

The estimated dynamic viscosity, μ_u, of urine at normal body temperature $T = 37$ °C, is $0.6531 < \mu_u < 0.7978$ (mPa·s) and falls within the range of that of water, $0.548 < \mu_w < 1.002$ (mPa·s). Experiments have also shown that μ_u is sensitive to the variation of the fraction of solutes, i.e.

$$\mu_u = \begin{cases} 0.9 \exp\left(1.5\dfrac{[...]}{1 - [...]}\right) & \text{for}[...] < 0.5 \\ 1.42 \exp\left(1.25\dfrac{[...]}{1 - [...]}\right) & \text{for } [...] > 0.5. \end{cases} \tag{8.16}$$

Although the dynamic viscosity of fluids changes with temperature, the human body operates within 35–40.5 °C . The decrease or increase in T above the physiological limits is seen in extreme pathological conditions—severe hypothermia, malignant hyperthermia—and is incompatible with life. Additionally, the kinematic viscosity, ν_u, ($\nu_u = \mu_u/\rho_u$), varies between $0.6518 < \nu_u < 0.795$ and $\nu_u \sim \nu_w$. It follows that urine can be treated as a viscous incompressible Newtonian fluid.

To close the problem of the urine-bladder–brain axis, the equations of motion of the bioshell, complemented by auxiliary relationships—constitutive relations for the tissue and its myoelectrical activity—along with the BBA model, initial and boundary conditions should be added. As mentioned previously, the model is multicompartmental, multiscale, large and hybrid and to execute it numerically is a formidable but possible task!

Different nuclei and centres in the brain including the medullary raphe nuclei, the locus coeruleus, the brain stem, Barrington's nucleus, the caudal and preoptic hypothalamus and the cerebral cortex, are all involved in regulating the lower-urinary-tract function. At the time of writing, the exact anatomical connections, mechanisms and pathways engaged in their signalling and regulation are unclear. For example: during normal bladder filling (distension), primary micturition centre neurons and the parasympathetic pathway are turned off, but at a critical level of sensitisation and visceral hypersensitivity, the afferent signalling switches them to maximal activity; the descending central pathways modulate the transmission of nociceptive information at the spinal cord level and lead to somatization. Existing noninvasive investigative techniques are also limited. The use of imaging studies such as positron-emission tomography and magnetic resonance along with functional magnetic resonance imaging in urology are still in their infancy. Little quantitative information is available regarding the CNS role in functional bladder disorders (Leue *et al* 2016). Unravelling the mysteries of the brain–bladder interaction will throw light on the intricate mechanisms of urine flow dynamics and the adaptive physiological and pathophysiological responses of the UB, to provide novel treatment targets and therapies for patients suffering from detrusor hyperreflexivity, overactive bladder, stress urinary incontinence and postsurgical dysfunction, among others.

8.3 Implementations in engineering

Biological and engineered systems are redesigned and replaced when they do not perform, albeit the former by evolution and the latter through customer dissatisfaction (Fletcher 2016). Over the past three decades the main interest in mechanobiology has fallen on the mechano-regulated morphogenesis and pathogenesis of G&R of: the heart, the abdominal aorta and cerebral arteries, the brain, wound healing, skin grafts and tumour growth (Dallon *et al* 2000, Tomasek *et al* 2002, Watton *et al* 2004, Gasser *et al* 2006, Menzel 2005, 2007, Baek *et al* 2006, 2007, Figueroa *et al* 2008, Valentin *et al* 2008, Goriely and Vandiver 2010, Karšaj *et al* 2010, Grytz *et al* 2011, 2012, Murphy 2011, Murphy *et al* 2012, Buganza-Tepole and Kuhl 2013, Valentín *et al* 2013, Bayly *et al* 2013, 2014, Valero *et al* 2015, Grytsan *et al* 2015, Aparício *et al* 2016, Lindquist Liljeqvist *et al* 2016, Li and Luo 2016, Koppenol and Vermolen 2017, Koppenol *et al* 2017, Virag *et al* 2017, Braeu *et al* 2017, 2019). For each problem, the specific evolution and constitutive relations has to be constructed and/or chosen. Thus, growth kinematics of: the brain or a tumour is assumed to be isotropic, $\mathbf{F}_g = \vartheta\mathbf{I}$; skeletal and cardiac muscle, transversely isotropic, $\mathbf{F}_g = \mathbf{I} + (\vartheta - 1)\boldsymbol{n} \otimes \boldsymbol{n}$ ($\boldsymbol{n}$ is a given direction); the skin, orthogonal $\mathbf{F}_g = \vartheta\mathbf{I} + (1 - \vartheta)\boldsymbol{n} \otimes \boldsymbol{n}$, or arbitrary (Menzel and Kuhl 2012). In almost all cases, a growth parameter ϑ is considered to be purely morphogenetic with an evolution independent of physical factors. Despite their computational practicality and effectiveness, e.g. the bioengineering of heart valves and vascular grafts, existing growth models have not been widely embraced by the community of biomedics. A healthy dose of rational scepticism and doubt regarding their credibility and accuracy are substantiated by the fact that neither address questions of genetic expressions, intracellular signalling, cell and tissue interactions, and, most importantly, the multicompartmental hierarchical integration of processes in a holistic system. Besides, the overall need for bioengineering input in soft tissue G&R is becoming very limited owing to the emergence and rapid development of stem cell and 3-dimensional printing technologies.

The main beneficiary in this milieu, however, is engineering science. Results of spatiotemporal modelling of complex biological systems can be crucial in the design of new materials with preprogrammed mechanical properties that replicate the functionality of their prototypes and perform at high levels of efficiency in various environments. Thus, fiberglass (an analogue of collagen and elastin fibres embedded in the ground matrix substance) and honeycomb (a replica of a bee's nest) composite materials have become indispensable in space, aviation, naval and vehicle industries for their strength, deformability, light-weight and space-economy. The variability in shapes (the regular hexagonal, chiral, triangular, star-shaped, etc), geometry (the core density, size, thickness of the wall, angles between ribbons), and constructive materials provided by nature, continues to inspire engineers in their search for extra-light and extra-durable materials.

References

Aparício P, Thompson M S and Watton P N 2016 A novel chemo-mechano-biological model of arterial tissue growth and remodeling *J. Biomech.* **49** 2321–30

Babtie A C and Stumpf M P H 2017 How to deal with parameters for whole-cell modelling *J. R. Soc. Interface* **14** 201170237

Baek S, Rajagopal K R and Humphrey J D 2006 A theoretical model of enlarging intracranial fusiform aneurysms *ASME J. Biomech. Eng.* **128** 142–9

Baek S, Valentin A and Humphrey J D 2007 Biochemomechanics of cerebral vasospasm and its resolution: II. Constitutive relations and model simulations *Ann. Biomed. Eng.* **35** 1498–509

Bayly P, Okamoto R, Xu G, Shi Y and Taber L 2013 A cortical folding model incorporating stress-dependent growth explains gyral wavelengths and stress patterns in the developing brain *Phys. Biol.* **10** 016005

Bayly P V, Taber L A and Kroenke C D 2014 Mechanical forces in cerebral cortical folding: a review of measurements and models *J. Mech. Behav. Biomed. Mater.* **29** 568–81

Birder L A, Kanai A J, DeGroat W C, Kiss S, Nealan M L, Burke N E, Dineley K E, Watkins S, Reynolds I J and Caterina M J 2001 Vanilloid receptor expression suggests a sensory role for urinary bladder epithelial cells *Proc. Natl Acad. Sci. USA* **98** 13396–401

Black J 1993 Drugs from emasculated hormones: the principles of syntopic antagonism *Nobel. Lect. Physiol. Med.* 1981–1990 ed T Frängsmyr (Singapore: World Scientific)

Braeu F A, Aydin R C and Cyron C J 2019 Anisotropic stiffness and tensional homeostasis induce a natural anisotropy of volumetric growth and remodeling in soft biological tissues *Biomech. Model. Mechanobiol.* **18** 327–45

Braeu F A, Seitz A, Aydin R C and Cyron C J 2017 Homogenized constrained mixture models for anisotropic volumetric growth and remodeling *Biomech. Model. Mechanobiol.* **16** 889–906

Buganza-Tepole A and Kuhl E 2013 Systems-based approaches toward wound healing *Pediatr. Res.* **73** 553–63

Cochran W G 1961 The role of mathematics in the medical sciences *New. Eng. J. Med.* July **27** 176

Dallon J, Sherratt J and Maini P 2000 Biological implications of a discrete mathematical model for collagen deposition and alignment in dermal wound repair *IMA J. Math. Appl. Med. Biol.* **17** 379–93

de Groat W C 2004 The urothelium in overactive bladder: passive bystander or active participant? *Urology* **64** 7–11

Figueroa C A, Baek S, Taylor C A and Humphrey J D 2008 A computational framework for coupled solid–fluid–growth mechanics in cardiovascular simulations *Comput. Meth. Appl. Mech. Eng.* **198** 3514–23

Fletcher D A 2016 Bottom-up biology: harnessing engineering to understand nature *Dev. Cell* **38** 587–9

Gasser T C, Ogden R W and Holzapfel G A 2006 Hyperelastic modelling of arterial layers with distributed collagen fibre orientations *J. R. Soc. Interface* **3** 15–35

Gerstner W, Sprekeler H and Deco G 2012 Theory and simulation in neuroscience *Science* **338** 60–5

Goriely A and Vandiver R 2010 On the mechanical stability of growing arteries *IMA J. Appl. Math.* **75** 549–70

Grytsan A, Watton P N and Holzapfel G A 2015 A thick-walled fluid–solid-growth model of abdominal aortic aneurysm evolution: application to a patient-specific geometry *J. Biomech. Eng.* **137** 031008

Grytz R, Meschke G and Jonas J B 2011 The collagen fibril architecture in the lamina cribrosa and peripapillary sclera predicted by a computational remodeling approach *Biomech. Model. Mechanobiol.* **10** 371–82

Grytz R, Sigal I A, Ruberti J W, Meschke G and Downs J C 2012 Lamina cribrosa thickening in early glaucoma predicted by a microstructure motivated growth and remodeling approach *Mech. Mater.* **44** 99–109

Gunawardena J 2014 Models in biology: 'accurate descriptions of our pathetic thinking' *BMC Biol.* **12** 29

Hasenauer J, Jagiella N, Hross S and Theis F J 2015 Data-driven modelling in biological multi-scale processes *J. Coupled. Syst. Multiscale. Dyn.* **3** 101–21

Huppert A and Kartiel G 2013 Mathematical modelling and prediction in infectious disease epidemiology *Clin. Microbiol. Infect.* **19** 999–1005

Karšaj I, Sorić J and Humphrey J D 2010 A 3-D framework for arterial growth and remodeling in response to altered hemodynamics *Int. J. Eng. Sci.* **48** 1357–72

Keener J and Sneyd J 2009 *Mathematical Physiology, I: Cellular Physiology* (Berlin: Springer)

Koppenol D C and Vermolen F J 2017 Biomedical implications from a morphoelastic continuum model for the simulation of contracture formation in skin grafts that cover excised burns *Biomech. Model. Mechanobiol.* **16** 1187–206

Koppenol D C, Vermolen F J, Niessen F B, van Zuijlen P P M and Vuik K 2017 A biomechanical mathematical model for the collagen bundle distribution-dependent contraction and sub-sequent retraction of healing dermal wounds *Biomech. Model. Mechanobiol.* **16** 345–61

Le Novère N 2015 Quantitative and logic modelling of molecular and gene networks *Nat. Rev. Genet.* **16** 146–58

Leue C, Kruimel J, Vrijens D, Masclee A, van Os J and van Koeveringe G 2016 Functional urological disorders: a sensitized defence response in the bladder–gut–brain axis *Nat. Rev. Urol.*

Li W and Luo X Y 2016 An invariant-based damage model for human and animal skins *Ann. Biomed. Eng.* **44** 3109–22

Liepe J, Fillipi S, Komorowski M and Stumpf M P H 2013 Maximising the information content of experiments in systems biology *PLoS Comput. Biol.* **9** e1002888

Lindquist Liljeqvist M, Hultgren R, Gasser T C and Roy J 2016 Volume growth of abdominal aortic aneurysms correlates with baseline volume and increasing finite element analysis-derived rupture risk *J. Vasc. Surg.* **63** 1434–42.e3

May R M 2004 Uses and abuses of mathematics in biology *Science* **303** 790–3

Menzel A 2005 Modeling of anisotropic growth in biological tissue. A new approach and computational aspects *Biomech. Model. Mechanobiol.* **3** 147–71

Menzel A 2007 A fiber reorientation model for orthotropic mutiplicative growth. Configurational driving stresses, kinematic-based reorientation, and algorithmic aspects *Biomechan. Model Mechanobiol.* **6** 303–20

Menzel A and Kuhl E 2012 Frontiers in growth and remodeling *Mech. Res. Commun.* **42** 1–14

Miftahof R and Nam H G 2013 *Biomechanics of the Human Urinary Bladder* (Berlin: Springer)

Murphy K 2011 Mathematical investigation of the interactions between the inflammatory response and mechanical aspects of dermal wound repair *PhD Thesis* Queensland Univ. Tech, Australia

Murphy K E, Hall C L, Maini P K, McCue S W and McElwain D L S 2012 A fibrocontractive mechanochemical model of dermal wound closure incorporating realistic growth factor kinetics *Bull. Math. Biol.* **74** 1143–70

Murray J D 2012 Vignettes from the field of mathematical biology: the application of mathematics to biology and medicine *Interf. Focuse* **2** 397–406

Putnam D F 1971 Composition and concentrative properties of human urine *Contr. Rep.* Number: NASA CR-1802

Sedov L I 1976 *Mechanics of Continuum* (Moskva: Nauka)

Tomasek J J, Gabbiani G, Hinz B, Chaponnier C and Brown R A 2002 Myofibroblasts and mechanoregulation of connective tissue remodelling *Nat. Rev. Mol. Cell Biol.* **3** 349–63

Tziannaros M and Smith F T 2012 Numerical and analytical study of bladder-collapse flow *Int. J. Diff. Eq.* **2012** 453467

Valentin A, Cardamone L, Baek S and Humphrey J D 2008 Complementary vaso-activity and matrix turnover in arterial adaptations to altered flow and pressure *J. R. Soc. Interface* **6** 293–306

Valentín A, Humphrey J and Holzapfel G A 2013 A finite element-based constrained mixture implementation for arterial growth, remodeling, and adaptation: Theory and numerical verification *Int. J. Numer. Methods Biomed. Eng.* **29** 822–49

Valero C, Javierre E, Garcia-Aznar J M, Menzel A and Gómez Benito M J 2015 Challenges in the modeling of wound healing mechanisms in soft biological tissues *Ann. Biomed. Eng.* **43** 1654–65

Villaverde A F and Banga J R 2013 Reverse engineering and identification in systems biology: strategies, perspectives and challenges *J. R. Soc. Interface* **11** 20130505

Virag L, Wilson J S, Humphrey J D and Karsaj I 2017 Potential biomechanical roles of risk factors in the evolution of thrombus-laden abdominal aortic aneurysms *Int. J. Numer. Meth. Biomed. Eng.* **33** e2893

Walpole J, Papin J A and Pierce S M 2013 Mutiscale computational models of complex biological systems *Ann. Rev. Biomed. Eng.* **15** 137–54

Watton P, Hill N and Heil M 2004 A mathematical model for the growth of the abdominal aortic aneurysm *Biomech. Model. Mechanobiol.* **3** 98–113

Wiseman O J, Brady C M, Hussain I F, Dasgupta P, Watt H and Fowler C J 2002 The ultrastructure of bladder lamina propria nerves in healthy subjects and patients with detrusor hyperreflexia *J. Urol.* **168** 2040–5

Zavala E, Wedgwood K C A, Voliotis M, Tabak J, Spiga F, Lightman S L and Tsaneva-Atanasova K 2019 Mathematical modelling in endocrine systems *Trends Endocrin. Metab.* **30** 244–57

IOP Publishing

Roustem N Miftahof and Christian J Cyron

Appendix A

A.1 Governing equations in orthogonal Cartesian coordinates

Let a soft shell be associated with an orthogonal Cartesian coordinate system x_1, x_2, x_3

$$\begin{aligned} x_1 &= x_1(\alpha_1, \alpha_2, t) \\ x_2 &= x_2(\alpha_1, \alpha_2, t), \\ x_3 &= x_3(\alpha_1, \alpha_2, t). \end{aligned} \tag{A.1}$$

The position vector of point $M(\alpha_1, \alpha_2) \in S$ and its derivatives are given by

$$\bar{r} = \bar{i}_1 x_1 + \bar{i}_2 x_2 + \bar{i}_3 x_3 = \sum_{k=1}^{3} x_k \bar{i}_k, \tag{A.2}$$

$$\bar{r}_i = \frac{\partial \bar{r}}{\partial \alpha_i} = \sum_{k=1}^{3} \frac{\partial x_k}{\partial \alpha_i} \bar{i}_k = \sum_{k=1}^{3} \bar{r}_{ik} \bar{i}_i, \quad (i = 1,2) \tag{A.3}$$

where $\bar{r}_{ik} = \partial x_k / \partial \alpha_i$ is the projection of the ith basis vector on the x_1, x_2, x_3 axes. Decomposing the unit vectors $\bar{e}_i = \bar{r}_i / |\bar{r}_i|$ along the base $\{\bar{i}_1, \bar{i}_2, \bar{i}_3\}$, we get

$$\bar{e}_i = \sum_{k=1}^{3} l_{ik} \bar{i}_k = \frac{\bar{r}_i}{\sqrt{a_{ii}^*}} = \frac{\bar{r}_{ik}}{\sqrt{a_{ii}^*}} \bar{i}_k = l_{ik} \bar{i}_k, \tag{A.4}$$

where l_{ik} are the direction cosines defined by

$$l_{ik} := \cos(\bar{e}_i, \bar{i}_k) = \bar{r}_{ik} / \sqrt{a_{ii}^*}. \tag{A.5}$$

The vector $\bar{m}$ normal to $\bar{e}_i$ ($\bar{m} \perp \bar{e}_i$) is given by

$$\bar{m} = (\bar{e}_1 \times \bar{e}_2) \sqrt{a_{11}^* a_{22}^*} / \sqrt{a^*}. \tag{A.6}$$

doi:10.1088/978-0-7503-3567-6ch9

With the help of equations (A.1), (A.3), the direction cosines $l_{3k} = \cos(\bar{m}, \bar{i}_k)$ are found to be

$$\begin{aligned} l_{31} &= (l_{12}l_{23} - l_{13}l_{22})\sqrt{a_{11}^* a_{22}^*}/\sqrt{a^*}, \\ l_{32} &= (l_{13}l_{21} - l_{11}l_{23})\sqrt{a_{11}^* a_{22}^*}/\sqrt{a^*}, \\ l_{33} &= (l_{11}l_{22} - l_{12}l_{21})\sqrt{a_{11}^* a_{22}^*}/\sqrt{a^*}. \end{aligned} \tag{A.7}$$

Scalar products $\bar{e}_i\bar{e}_k$ yield

$$\begin{aligned} \bar{e}_1\bar{e}_2 &= l_{11}l_{21} + l_{12}l_{22} + l_{13}l_{23} = \cos\chi^* = a_{12}^*/\sqrt{a_{11}^* a_{22}^*} \\ \bar{e}_i\bar{e}_2 &= l_{i1}l_{31} + l_{i2}l_{32} + l_{i3}l_{33} = 0, \\ \bar{e}_k\bar{e}_k &= l_{k1}^2 + l_{k2}^2 + l_{k3}^2 = 1. \end{aligned} \tag{A.8}$$

Expanding $\bar{p}$, $\bar{f}$ in the direction of $\bar{e}_i$ and $\bar{i}_i$, respectively, we obtain

$$\bar{p} = \bar{e}_1 p_1 + \bar{e}_2 p_2 + \bar{m} p_3, \tag{A.9}$$

$$\bar{f} = \bar{i}_1 f_1 + \bar{i}_2 f_2 + \bar{i}_3 f_3. \tag{A.10}$$

Substituting equations (A.5), (A.7), (A.9), (A.10) in equation (3.11), the equation of motion of the soft shell takes the form

$$\begin{aligned} \rho\sqrt{a}\frac{d^2x_1}{dt^2} &= \frac{\partial}{\partial\alpha_1}\left[(T^{11}l_{11} + T^{12}l_{21})\sqrt{a_{22}^*}\right] + \frac{\partial}{\partial\alpha_2}\left[(T^{12}l_{11} + T^{22}l_{21})\sqrt{a_{11}^*}\right] + \\ &\quad + (p_1 l_{11} + p_2 l_{21})\sqrt{a^*} + p_3(l_{21}l_{23} - l_{13}l_{22})\sqrt{a_{11}^* a_{22}^*} + \rho f_1\sqrt{a} \\ \rho\sqrt{a}\frac{d^2x_2}{dt^2} &= \frac{\partial}{\partial\alpha_1}\left[(T^{11}l_{12} + T^{12}l_{22})\sqrt{a_{22}^*}\right] + \frac{\partial}{\partial\alpha_2}\left[(T^{12}l_{12} + T^{22}l_{22})\sqrt{a_{11}^*}\right] + \\ &\quad + (p_1 l_{12} + p_2 l_{22})\sqrt{a^*} + p_3(l_{13}l_{21} - l_{11}l_{23})\sqrt{a_{11}^* a_{22}^*} + \rho f_2\sqrt{a}, \\ \rho\sqrt{a}\frac{d^2x_3}{dt^2} &= \frac{\partial}{\partial\alpha_1}\left[(T^{11}l_{13} + T^{12}l_{23})\sqrt{a_{22}^*}\right] + \frac{\partial}{\partial\alpha_2}\left[(T^{12}l_{13} + T^{22}l_{23})\sqrt{a_{11}^*}\right] + \\ &\quad + (p_1 l_{13} + p_2 l_{23})\sqrt{a^*} + p_3(l_{11}l_{22} - l_{12}l_{21})\sqrt{a_{11}^* a_{22}^*} + \rho f_3\sqrt{a}. \end{aligned} \tag{A.11}$$

A.2 Governing equations in cylindrical coordinates

Let a soft shell be associated with a cylindrical coordinate system $\{r, \varphi, z\}$

$$\begin{aligned} r &= r(\alpha_1, \alpha_2, t) \\ \varphi &= \varphi(\alpha_1, \alpha_2, t), \\ z &= z(\alpha_1, \alpha_2, t). \end{aligned} \tag{A.12}$$

It is related to the orthogonal Cartesian coordinates $\{x_1, x_2, x_3\}$ as

$$x_1 = r\cos\varphi, \quad x_2 = r\sin\varphi, \quad x_3 = z.$$

The position vector $\bar{r}$ of point $M(r, \varphi, z) \in S$ is given by

$$\bar{r} = r\bar{k}_1 + z\bar{k}_3, \tag{A.13}$$

where

$$\bar{k}_1 = \bar{i}_1 \cos\varphi + \bar{i}_2 \sin\varphi, \quad \bar{k}_2 = -\bar{i}_1 \sin\varphi + \bar{i}_2 \cos\varphi. \tag{A.14}$$

Differentiating equation (A.13) with respect to α_i with the help of equation (A.14), we find

$$\begin{aligned} \bar{r}_i = \frac{\partial \bar{r}}{\partial \alpha_i} &= \frac{\partial r}{\partial \alpha_i}\bar{k}_1 + r\frac{\partial \bar{k}_1}{\partial \alpha_i} + \frac{\partial z}{\partial \alpha_i}\bar{k}_3 \\ &= \frac{\partial r}{\partial \alpha_i}\bar{k}_1 + r\frac{\partial \varphi}{\partial \alpha_i}\bar{k}_2 + \frac{\partial z}{\partial \alpha_i}\bar{k}_3, \end{aligned} \tag{A.15}$$

where

$$\begin{aligned} \frac{\partial \bar{k}_1}{\partial \alpha_i} &= \frac{\partial}{\partial \alpha_i}(\bar{i}_1 \cos\varphi + \bar{i}_2 \sin\varphi) \\ &= \frac{\partial \varphi}{\partial \alpha_i}(-\bar{i}_1 \sin\varphi + \bar{i}_2 \cos\varphi) = \frac{\partial \varphi}{\partial \alpha_i}\bar{k}_2. \end{aligned} \tag{A.16}$$

Projections of $\bar{r}_i$ in the direction of the r, φ, and z axes are given by

$$\bar{r}_{ir} = \frac{\partial r}{\partial \alpha_i}, \quad \bar{r}_{i\varphi} = \frac{\partial \varphi}{\partial \alpha_i}, \quad \bar{r}_{iz} = \frac{\partial z}{\partial \alpha_i}.$$

Hence,

$$\bar{r}_i = \bar{r}_{ik}\bar{k}_1 + \bar{r}_{i\varphi}\bar{k}_2 + \bar{r}_{iz}\bar{k}_3. \tag{A.17}$$

Decomposing $\bar{e}_i$ along the base $\left\{\bar{k}_1, \bar{k}_2, \bar{k}_3\right\}$, we find

$$\bar{e}_i = l_{ir}\bar{k}_1 + l_{i\varphi}\bar{k}_2 + l_{iz}\bar{k}_3. \tag{A.18}$$

Here the direction cosines $l_{ij}(i = 1, 2, 3; j = r, \varphi, z)$ are given by

$$\begin{aligned} &\bar{e}_1\bar{e}_2 = l_{1r}l_{2r} + l_{1\varphi}l_{2\varphi} + l_{1z}l_{2z} = \cos\chi^* = a_{12}^*/\sqrt{a_{11}^*a_{22}^*} \\ &\bar{e}_i\bar{e}_3 = l_{ir}l_{3r} + l_{i\varphi}l_{3\varphi} + l_{iz}l_{3z} = 0, \\ &\bar{e}_n\bar{e}_n = l_{n1}^2 + l_{n2}^2 + l_{n3}^2 = 1, \quad (n = 1,2,3). \end{aligned} \tag{A.19}$$

Expanding the resultant of external, $\bar{p}$, and mass, $\bar{f}$, forces in the direction of unit vectors $\bar{e}_i$, we get

$$\bar{p} = \bar{e}_1 p_1 + \bar{e}_2 p_2 + \bar{e}_3 p_3, \tag{A.20}$$

$$\bar{f} = \bar{k}_1 f_r + \bar{k}_2 f_\varphi + \bar{k}_3 f_z. \tag{A.21}$$

The vector of acceleration $\bar{a}(a_r, a_\varphi, a_z)$ in cylindrical coordinates is given by

$$a_r = -\frac{d^2r}{dt^2} - r\left(\frac{d\varphi}{dt}\right)^2, \quad a_\varphi = -r\frac{d^2\varphi}{dt^2} + 2r\frac{dr}{dt}\frac{d\varphi}{dt}, \quad a_z = -\frac{d^2z}{dt^2}. \tag{A.22}$$

Substituting equations (A.18), (A.20)–(A.22) in equation (3.11), the equations of motion of the soft shell in cylindrical coordinates take the form

$$\begin{aligned}
\rho\sqrt{a}\left(\frac{d^2r}{dt^2} - r\left(\frac{d\varphi}{dt}\right)^2\right) &= \frac{\partial}{\partial\alpha_1}\left[(T^{1r}l_{1r} + T^{2r}l_{2r})\sqrt{a_{12}^*}\right] \\
&+ \frac{\partial}{\partial\alpha_2}\left[(T^{1r}l_{1r} + T^{2r}l_{2r})\sqrt{a_{11}^*}\right] + (p_1 l_{1r} + p_2 l_{2r})\sqrt{a^*} \\
&+ p_3(l_{2\varphi}l_{2z} - l_{1z}l_{2\varphi})\sqrt{a_{11}^* a_{22}^*} + \rho f_r\sqrt{a} \\
\rho\sqrt{a}\left(r\frac{d^2\varphi}{dt^2} + 2r\frac{dr}{dt}\frac{d\varphi}{dt}\right) &= \frac{\partial}{\partial\alpha_1}\left[(T^{1\varphi}l_{1\varphi} + T^{1\varphi}l_{2\varphi})\sqrt{a_{22}^*}\right] \\
&+ \frac{\partial}{\partial\alpha_2}\left[(T^{1\varphi}l_{1\varphi} + T^{2\varphi}l_{2\varphi})\sqrt{a_{11}^*}\right] + (p_1 l_{1\varphi} + p_2 l_{2\varphi})\sqrt{a^*} \\
&+ p_3(l_{1z}l_{2r} - l_{1r}l_{2z})\sqrt{a_{11}^* a_{22}^*} + \rho f_\varphi\sqrt{a} \\
\rho\sqrt{a}\frac{d^2z}{dt^2} &= \frac{\partial}{\partial\alpha_1}\left[(T^{1z}l_{1z} + T^{2z}l_{2z})\sqrt{a_{22}^*}\right] + \frac{\partial}{\partial\alpha_2}\left[(T^{1z}l_{1z} + T^{2z}l_{2z})\sqrt{a_{11}^*}\right] \\
&+ (p_1 l_{1z} + p_2 l_{2z})\sqrt{a^*} + p_3(l_{1r}l_{2\varphi} - l_{1\varphi}l_{2r})\sqrt{a_{11}^* a_{22}^*} + \rho f_z\sqrt{a}.
\end{aligned} \tag{A.23}$$

Appendix B

B.1 cAMP-dependent pathway

The production of cAMP is controlled by a number of adenylate cyclase (AC) enzymes which are dually regulated by a family of G proteins, forskolin, and other class-specific substrates. The diversity and expression of AC isoforms (I–IX) provide a mechanism for integrating positively or negatively the responses to various ligands. Thus, (i) $AC_{V/VI}$ have a high affinity to $G_{\beta\gamma}$ subunits and are inhibited directly by low levels of Ca_i^{2+}, (ii) $AC_{I/III/VIII}$ are shown to be up-regulated by a calmodulin dependent protein kinase in response to an elevation in .., whilst (iii) $AC_{II/IV/IX}$ are stimulated in the presence of $G\alpha_s$ subunits. Although the exact distribution of specific types of isoforms in the hypothalamus is not known, it is obvious that the very same stimulus may trigger different physiological responses depending not only on the type of receptors involved but also on the type of AC to which they are coupled.

The proposed structure of ACs includes a short amino-terminal and two cytoplasmic domains. The latter are separated by two hydrophobic regions in the form of six transmembrane helices. The catalytic core of the enzyme consists of a pseudosymmetric heterodimer composed of two highly conserved portions of the cytoplasmic domains. It binds one molecule of $G\alpha_s$ which catalysers the conversion of ATP into cAMP. The latter influences a wide range of effects including: (i) the increase in Ca^{2+} channel conductance, (ii) the activation of PkA, and (iii) the production of cytokines.

Protein kinase A is a holoenzyme containing two regulatory and two catalytic subunits. The binding of cAMP to the regulatory subunits causes the release of the catalytic subunits and the transfer of ATP terminal phosphates to the effector molecule. PkA promotes the inhibition of PLC, intracellular Ca^{2+} entry to activate BK_{Ca} channels and calcium pumps. Additionally, it decreases the affinity of myosin light chain kinase for the calcium-calmodulin complex and, as a result, relaxes smooth muscle. PkA activity is controlled entirely by cAMP and remains catalytically inactive under low levels of the enzyme.

doi:10.1088/978-0-7503-3567-6ch10

The concentration of cAMP is regulated both by AC and phosphodiesterases (PDEs) that degrade it to 5′-AMP. The hydrolytic activity of PDEs is determined by the catalytic domain and conserved areas of amino acids specific to each family. Thus, PDE4, 7 and 8 hydrolyse cAMP, whilst PDE5, 6 and 9 - cGMP and the rest demonstrate dual specificity to cAMP and cGMP (Wheeler *et al* 2005). In humans, PDE 1, 2, 3 and 5 enzymes are specific for cGMP, and PDE4 for cAMP. Another mechanism that reduces the production of cAMP is the activation of $G\alpha_{iq/11}$ proteins which directly inhibit AC through the MAPK signalling mechanism.

B.2 PLC-dependent pathway

Four β, two γ, four δ, and ε isoforms of PLC enzymes have been identified. PLCβ members are regulated mainly by Ca^{2+}, but respond differently to G proteins. PLCβ1 and PLCβ4 are sensitive to $G\alpha_{q/11}$, whereas PLCβ2 and PLCβ3 can be activated by $G\alpha_{q/11}$ and $G_{\beta\gamma}$ subunits. Without exception, their activation leads to the break down of inositide-4,5-biphosphate (PIP_2) with the production of inositol-1,4,5-triphosphate (IP_3) and 1,2-diacylglycerol (DAG). IP_3 is a highly soluble structure that diffuses quickly through the cytosol towards the endoplasmic reticulum where it binds to IP_3 surface receptors (IP_3R) to trigger the mobilisation of stored Ca^{2+}_{ER}.

Diacylglycerol consists of two fatty acid chains covalently bonded to a glycerol molecule. Compared to IP_3, it is hydrophobic and remains bound to the plasma membrane. It has a number of functions including the activation of PkC and a subfamily of TRPC cation channels. Together with acyl-CoA, it is converted to triacylglycerol by the addition of a third fatty acid to its molecule. The reaction is catalysed by two distinct isoforms of diglyceride acyltransferases.

Protein kinase C comprises a family of eleven isoenzymes divided into three subfamilies—conventional ($\alpha,\beta_1,\beta_2,\gamma$), novel ($\delta,\varepsilon,\eta,\theta,\mu$), and atypical ($\zeta,\iota,\lambda$)—based on their second messenger requirements for activation. PKCs consist of a variable regulatory and a highly conserved catalytic domain, tethered together by a hinge region. The regulatory domain is divided in two subregions, C1 that binds to DAG and phorbol esters, and C2 that acts as a Ca^{2+} sensor. The latter is functional only in conventional PKCs. Upon activation, kinases are translocated to the plasma membrane where they remain active for a long period of time.

B.3 PI3K/PkB pathway

The PI3K/PkB pathway mediates a wide range of cell functions including cell growth, proliferation, differentiation, protein synthesis, and intracellular trafficking among many others. The key players are: (i) PI3Ks—a family of plasma membrane attached to intracellular signal transducer enzymes, and (ii) PkB proteins that reside freely in the cytosol in a relaxed form. Four different classes of PI3Ks (I–IV) have been identified based on their primary structure and function. Class I PI3Ks are involved in lipid phosphorylation; class II catalyse the production of phosphatidylinositol 3-phosphate (PI3P) from phosphatidylinositol (PI), and also phosphatidylinositol (3,4)-bisphosphate from phosphatidylinositol 4-phosphate; class III kinases has a primary function in the trafficking of proteins and vesicles; the role of class IV kinases remains unclear. Structurally, PI3Ks (I) are heterodimers composed of a regulatory (p85) and a catalytic (p110) subunit. Upon stimulation of GH- and/or IGF-1 receptors, the p85 subunit binds at its Src homology 2 domain to phosphorylated tyrosine residues and subsequently recruits the catalytic subunit. This fully activates the PI3K which catalysers the conversion of phosphatidylinositol (4,5)-bisphosphate (PIP_2) into phosphatidylinositol (3,4,5)-trisphosphate (PIP_3). The p110 subunit as well as growth factor receptor-bound protein 2- associated scaffold protein can be activated directly by IRS-1. Additionally, PI3Ks can be stimulated by $G_{\beta\gamma}$ protein dimers, Ras or via the Gα-Shc-dependent integrin signalling pathway. The main downregulatory protein of PI3Ks is phosphatase and tensin homologue.

Three isoforms (I–III) of PkB have been isolated but only $PkB_{I/II}$ are involved in protein synthesis and insulin signalling. PkB possesses a pleckstrin homology domain which binds with high affinity to phosphoinositides, either PIP_3 or PIP_2, attached to the cell membrane. Once positioned on the membrane, a small amount of PkB is phosphorylated by phosphoinositide dependent kinase 1. Active PkB translocates to the nucleus where it stimulates ribosomal protein S6 kinase polypeptide 1 (S6K1). This subsequently phosphorylates ribosomal protein S6, promoting protein synthesis and cellular proliferation. The non-phosphorylated PkB is ubiquitinated and degraded by proteasome.

B.4 cGMP-dependent PkG

Protein kinase G phosphorylates a number of biological targets and is implicated in the regulation of cell division and nucleic acid synthesis, among others. Two isoforms (I, II) of PkG are identified in the dorsal root ganglia and brain neurones. In the cell, PkG-Is are localised primarily in the cytoplasm whilst PkG-IIs are anchored to the plasma membrane. Structurally, PkGs are homodimers of two identical subunits, each of which is composed of three domains: (i) an N-terminal domain mediating homodimerization, the suppression of kinase activity in the absence of cGMP and the interactions with other proteins; (ii) a regulatory domain inducing a disinhibitory conformational change of the catalytic core and promoting the phosphorylation of substrate proteins, and (iii) a kinase domain catalysering the phosphate transfer from ATP to the hydroxyl group of a serine/threonine side chain of the target protein. Once activated by cGMP, the enzyme hyperpolarizes the cell membrane by increasing the permeability of BK_{Ca} channels and deactivating PLC, thus negatively affecting the release of Ca^{2+} from the ER.

B.5 RAS/MAPK pathway

Membrane bound Ras-guanine nucleotide proteins belong to the class of GTPases. They serve as key transducers of receptor tyrosine kinase signalling and regulate several cellular processes including mitogenesis, apoptosis and DNA-damage repair. The proteins exist in inactive (Ras-GDP-bound) and active (Ras-GTP-bound) forms. Although Ras has similar affinities for GDP and GTP, a low GDP/GTP exchange rate between the two in the absence of stimulants, favours the Ras-GDP state in unexcited cells. The main trigger in Ras activation is guanine nucleotide exchange factors, e.g., $SOS_{1/2}$, the scaffold protein. These facilitate conformational changes of the Ras-GDP complex to its active GTP-bound state by the release of nucleotide (Cullis *et al* 2014). Ras binds to SOS at two distinct sites and acts both as a substrate and an allosteric activator. As a substrate, it initiates the recruitment of SOS to the plasma membrane, whilst when acting at the allosteric site it promotes the dissociation of GDP from Ras. Conversion to the inactive form is mediated by GTPase proteins.

The downstream involvement of specific intracellular signalling pathways by activated Ras, depends entirely on the location of proteins in the cell: caveolin-positive/negative regions of the plasma membrane, endomembranes of the endoplasmic reticulum, the Golgi complex, mitochondria, and/or endosomes (Prior and Hancock 2012). The Ras/MAPK pathway is ubiquitous in cells and is involved in cell growth, division and synchronisation of cellular functions. Mitogen-activated protein kinases belong to a family of silk proteins/threonine kinases comprised of ERK-1/2 and 5 (extracellular-signal-regulated); JNKs (Jun amino-terminal) and p38/SAPKs (stress-activated protein) kinases. Each is initiated by specific extracellular cues leading to the activation of particular MAPK kinases (Erickson *et al* 2018, Morrison 2012). These sequentially phosphorylate transcription factors, protein kinases, phospholipases and cytoskeletal proteins, as well as regulate a wide range of physiological responses. Thus, p38MAPK and JNK when activated under stress, inhibit cell growth, whilst ERK signalling promotes cell proliferation, development, and differentiation (McKay and Morrison 2007, Shaul and Seger 2007). The signalling downstream of the Ras/MAPK pathway contains multiple positive and negative feedback loops, protein re-localisation, intermediate complex formation and cross-talk between parallel signalling pathways. Thus, the stimulation of Sprouty (negative) proteins suppresses signalling from Ras to ERK, and activating instead the PI3K pathway, whilst the positive feedback from Ras to SOS may cause bistability in Ras-GTP regulation.

Inactivation of Ras/MAPK signalling involves several different mechanisms, including internalisation, ubiquitination, and degradation.

References

Cullis J and et al 2014 The RhoGEF GEF-H1 is required for oncogenic RAS signaling via KSR-1 *Canc. Cell* **25** 181–195

Erickson K E, Rukhlenko O S, Posner R G, Hlavaceka W S and Kholodenko B N 2018 New insights into RAS biology reinvigorate interest in mathematical modeling of RAS signalling *Sem. Canc. Biol.* **54** 162–73

McKay M M and Morrison D K 2007 Integrating signals from RTKs to ERK/MAPK *Oncogene* **26** 3113–3121

Morrison D K 2012 MAPK kinase pathways *Cold Spring Harb. Persp. Biol* **4** a011254

Prior I A and Hancock J F 2012 Ras tracking, localization and compartmentalized signalling *Semin. Cell. Dev. Biol.* **23** 145–153

Shaul Y D and Seger R 2007 The MEK/ERK cascade: From signaling specificity to diverse functions *Biochim. Biophys. Acta* **1773** 1213–1226

IOP Publishing

Roustem N Miftahof and Christian J Cyron

Appendix C

C.1 Endoplasmic reticulum and Ca^{2+} turnover

The major intracellular store for Ca^{2+} is the endoplasmic reticulum (ER). It is a folded continuation of the nuclear cell membrane interconnected by the network of enclosed flattened and folded sacs and tubes, known collectively as cisternae. The terminal cisternae face the cytosol being primarily involved in the processes of the absorption and release of calcium ions. Thus, part of the free extracellular Ca^{2+} that enters the cell through the transmembrane L- and T-type Ca^{2+} channels during depolarisation or after (the) release from the store, is actively transported into the ER by the membrane-bound Ca^{2+}-ATPase pump at a rate of ~10/s (Kekenes-Huskey *et al* 2012, Lytton *et al* 1992). Once in the lumen of cisternae, ions bind to calsequestrin, a major Ca^{2+}-buffering protein. This can also interact with the membrane-associated proteins junction, triadin and the ryanodine receptor (RyR) to form the core of the macromolecular complex that regulates Ca^{2+} release. However, much controversy surrounds their mechanism of action.

There are two Ca^{2+} release channels present on the ER: ryanodine and IP_3 receptors. Three different isoforms of RyRs (RyR_{1-3}) have been cloned to date. Although isoforms share approximately 65% sequence homology, they differ in: (i) tissue expression (the $RyR_{1/2}$ are mainly found in cardiac, skeletal and smooth muscle whilst the RyR_3 are in neuronal tissue in the brain), and (ii) responses to allosteric modulators including the dihydropyridine receptor, Ca^{2+}, Mg^{2+}, PkA, PkG, FK506 binding proteins and phosphatases, among others. Suffice is it to note that RyR_3s are the least studied so consequently little is known of their function.

The primary trigger for RyRs opening/closing is Ca^{2+}, the same ion that it permeates. Calcium 'positive' sensors within the receptor bind Ca^{2+} causing its conformational changes, pore opening, dissociation of Ca^{2+} from calsequestrin and ion release into the cytosol. As soon as the level of intracellular Ca^{2+} reaches the required concentration, Ca^{2+} binds to 'negative' sensors on the RyR promoting its inactivation and closure (van Petegem 2012, Gerhard Meissner 2017). By contrast, the IP_3R is modulated by inositol 1,4,5-trisphosphate, a diffusible messenger in the

doi:10.1088/978-0-7503-3567-6ch11

PLC pathway. Upon the binding of IP_3 and Ca^{2+} to the receptor, a tetrameric assembly is formed that opens the gate of the channel allowing Ca^{2+} to exit the ER (Mak and Foskett 2014). In both cases the release of ions is very rapid at (the) a rate of ~10^3 Ca^{2+}/s. However, the process is limited by the slower dynamics of second messenger turnover, PLC, which is only ~30/s (Waite 1985).

The first mathematical model to describe the kinetics of the IP_3 receptor/Ca^{2+}-turnover was developed by Keizer and De Young (1992) and De Young and Keizer (1992). The dynamics of the concentration of free cytosolic Ca^{2+} ($[Ca^{2+}]_{cyt}$) is given by

$$\frac{d[Ca^{2+}]_{cyt}}{dt} = J_{PMCa} - J_{ERCa}, \tag{C.1}$$

and J_{PMCa}, J_{ERCa}- the outward and influx of Ca^{2+} ions from the ER yield

$$\begin{aligned} J_{PMCa} &= c_1(v_1 P_{\text{open}} + v_2)\left([Ca^{2+}]_{ER} - [Ca^{2+}]_{cyt}\right) \\ J_{ERCa} &= \frac{v_3[Ca^{2+}]_{cyt}^2}{[Ca^{2+}]_{cyt}^2 + K_{PMCa}^2} \end{aligned} \tag{C.2}$$

Here, c_1, v_1, v_2, v_3, K_{PMCa} are adjustable empirical constants related to the fractional volume and biophysical properties of IP_3 / Ca^{2+} channels. Based on the assumptions that: (i) three equivalent and independent subunits are involved in channel conduction, (ii) each subunit has three binding sites, i.e., one for IP_3 and two for Ca^{2+} ions, (iii) Ca^{2+} directly affects the binding of IP_3 to the receptor, then the open probability of the channel at equilibrium, P_{open}, yields

$$P_{\text{open}} = \left[\frac{\bar{k}_2[Ca^{2+}]_{cyt}[IP_3]}{\left([Ca^{2+}]_{cyt}[IP_3] + \bar{k}_2[IP_3] + \bar{k}_1\bar{k}_2 + [Ca^{2+}]_{cyt}\bar{k}_3\right)\left([Ca^{2+}]_{cyt} + \bar{k}_4\right)}\right]^3 \tag{C.3}$$

where $\bar{k}_i(i = 1,4)$ are equilibrium rate constants, and the kinetics of inositol 1,4,5-trisphosphate is

$$\frac{d[IP_3]}{dt} = v_4\left(\frac{[Ca^{2+}]_{cyt} + (1 - \alpha)k_4}{\left([Ca^{2+}]_{cyt} + k_4\right)}\right) - I_r[IP_3]. \tag{C.4}$$

Here, v_4, k_4, are the maximum rate and the dissociation constant for Ca^{2+} stimulation of IP_3 production, respectively, and $\alpha(0 \leqslant \alpha \leqslant 1)$ is the parameter reflecting Ca^{2+} stimulation parameter of IP_3 production by PLC.

The nine-variable model *per se* reproduces the oscillatory behaviour of $[Ca^{2+}]_{cyt}$ as a result of IP_3 receptor/Ca^{2+}-channel activity with partial emptying and refilling of the store. Li and Rinzel (1994) reduced the model formulation to a two-variable system analogous in form to the Hodgkin-Huxley equations. Their analysis and results of numerical simulations have shown that the proposed model captures the important dynamic features of Ca^{2+}-induced Ca^{2+} release (CICR) from the ER, which is similar to that reported by Keizer and De Young (1992).

Currently, the general trend in modelling biological processes at all levels is reductionistic. The move is supported by recent large-scale data acquisition and an overwhelming growth in experimental results regarding the molecular mechanisms of signalling. Therefore highly specified, detailed models are being constructed to simulate behavioural patterns of interactions among the IP_3R, $[Ca^{2+}]_{ER}$, $[Ca^{2+}]_{cyt}$, extracellular Ca^{2+} and ATP, e.g. LeBeau *et al* (2000), van Goor *et al* (2000), Lavrentovich and Hemkin (2008), Li and Khadra (2008), Fletcher and Li (2009), Zeng *et al* (2009), Lanner *et al* (2010), Siekmann *et al* (2012), Pecze and Schwalle (2015), Dupont and Sneyd (2017), and Taylor (2017).

References

De Young G W and Keizer J 1992 A single pool IP3-receptor based model for agonist stimulated Ca^{2+} oscillations *Proc. Natl. Acad. Sci. USA* **89** 9895–9

Dupont G and Sneyd J 2017 Recent developments in models of calcium signalling *Curr. Opin. Sys. Biol.* **3** 15–22

Fletcher P A and Li Y-X 2009 An integrated model of electrical spiking, bursting, and calcium oscillations in GnRH neruons *Biophys. J.* **96** 4514–24

Gerhard Meissner G 2017 The structural basis of ryanodine receptor ion channel function *J. Gen. Physiol.* **149** 1065–89

Keizer J and De Young G W 1992 Two roles for Ca^{2+} in agonist stimulated Ca^{2+} oscillations *Biophys. J.* **61** 649–60

Kekenes-Huskey P M, Metzger V T, Grant B J and McCammon A J 2012 Calcium binding and allosteric signaling mechanisms for the sarcoplasmic reticulum Ca^{2+}-ATPase *Protein Sci.* **21** 1429–43

Lanner J T, Georgiou D K, Joshi A D and Hamilton S L 2010 Ryanodine receptors: structure, expression, molecular details, and function in calcium release *Cold Spring Harb. Perspect. Biol.* **2** a003996

Lavrentovich M and Hemkin S 2008 A mathematical model of spontaneous calcium (II) oscillations in astrocytes *J. Theor. Biol.* **251** 553–60

LeBeau A P, Van Goor F, Stojilkovic S S and Sherman A 2000 Modeling of membrane excitability in gonadotropin-releasing hormone-secreting hypothalamic neurons regulated by Ca^{2+}-mobilizing and adenylyl cyclase-coupled receptors *J. Neurosci.* **20** 9290–7

Li Y and Rinzel J 1994 Equatons for InsP3 receptor mediated $[Ca^{2+}]_i$ oscillations derived from a detailed kinetic model: a Hodgkin–Huxley like formalism *J. Theor. Biol.* **166** 461–73

Li Y-X and Khadra A 2008 Robust synchrony and rhythmogenesis in endocrine neurons via autocrine regulations *in vitro* and *in vivo Bull. Math. Biol.* **70** 2103–25

Lytton J, Wetlin M, Burk S E, Shull G E and MacLennan D H 1992 Functional comparisons between isoforms of the sarcoplasmic or endoplasmic reticulum family of calcium pumps *J. Biol. Chem.* **267** 14483–9

Mak D D and Foskett J K 2014 Inositol 1,4,5-trisphosphate receptors in the endoplasmic reticulum *Cell Calcium* **58** 67–78

Pecze L and Schwalle B 2015 Characterization and modelling of Ca^{2+} oscillations in mouse primary mesothelial cells *Biochim. Biophys. Acta* **1853** 632–45

Siekmann I, Wagner L E II, Yule D, Crampin E J and Sneyd J 2012 A kinetic model for type I and II IP3R accounting for mode changes *Biophys. J.* **103** 658–68

Taylor C 2017 Regulation of IP3 receptors by cyclic cAMP *Cell Calcium* **63** 48–52 van Goor F, LeBeau A P, Krsmanovic L Z, Sherman A, Catt K J and Stojilkovic S S 2000 Amplitude-dependent spike-broadening and enhanced Ca^{2+} signaling in GnRH-secreting neurons *Biophys. J.* **79** 1310–23

van Petegem S F 2012 Ryanodine receptors: structure and function *J. Biol. Chem.* **287** 31624–32

Waite M 1985 Approaches to the study of mammalian cellular phospholipases *J. Lipid Res.* **26** 1379–88

Zeng S, Li B, Zeng S and Chen S 2009 Simulation of sponatenous Ca^{2+} oscillations in astrocytes mediated by voltage-gated calcium channels *Biophys. J.* **97** 2429–37

Roustem N Miftahof and Christian J Cyron

Appendix D

Visceral tissues are considered to be a mixture of multiple constituents, $i \in I$, which includes a ground substance matrix with families of collagen and smooth muscle fibres. It is generally assumed that connective and smooth muscle tissue are subject to G&R whilst the matrix remains biologically inert.

Let a biocomposite in a given reference configuration Ω_0 under finite deformations and displacements attain a new state Ω_t (figure D.1): $\Omega_0 \overset{x}{\mapsto} \Omega_t$ in which the gradient of the mapping, $\boldsymbol{x}$, is

$$\boldsymbol{F} = \frac{\partial \boldsymbol{x}}{\partial \boldsymbol{X}}. \tag{D.1}$$

Here, $\boldsymbol{X} \in \Omega_0$, $x \in \Omega_t$ are position vectors of a material point in the initial (referential), Ω_0, and the deformed (current), Ω_t, configurations, respectively. These are related as

$$\boldsymbol{x}(\boldsymbol{X}, t) = \boldsymbol{X} + \boldsymbol{u}(\boldsymbol{X}, t) \tag{D.2}$$

where $\boldsymbol{u}(X, t)$is the displacement vector.

The associated Cauchy–Green, $\boldsymbol{C}$, and Green–Lagrange, $\boldsymbol{E}$, strain tensors are

$$\begin{aligned} \boldsymbol{C} &= \boldsymbol{F}^{\mathrm{T}}\boldsymbol{F} \\ \boldsymbol{E} &= \frac{1}{2}(\boldsymbol{F}^{\mathrm{T}}\boldsymbol{F} - \boldsymbol{I}) = \frac{1}{2}(\boldsymbol{C} - \boldsymbol{I}). \end{aligned} \tag{D.3}$$

Assuming further that all constituents of the biocomposite deform simultaneously, i.e. their deformation under loading is constrained, then the strain $\boldsymbol{E}$ and the deformation gradient $\boldsymbol{F}$ remain the same for all constituents. To reflect the individuality of G&R for each constituent i, in the mixture, the deformation gradient for each constituent can be given by

$$F = \boldsymbol{F}_e^i \boldsymbol{F}_{gr}^i. \tag{D.4}$$

doi:10.1088/978-0-7503-3567-6ch12

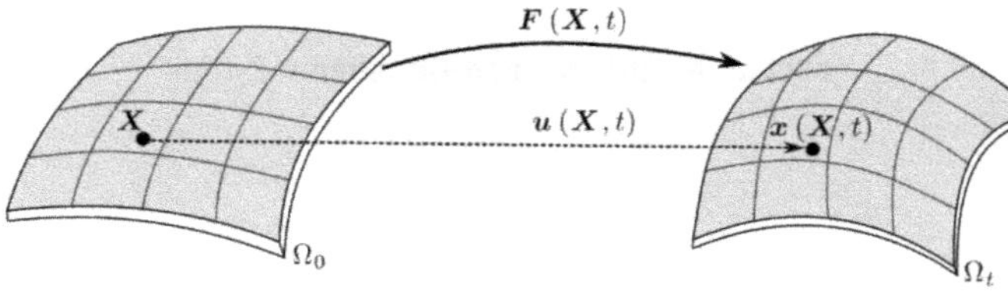

Figure D.1. Deformation of a biocontinuum.

Herein the superscript, *i*, indicates a constituent and subscripts *e* and *gr* refer to elasticity and G&R, respectively. Thus $\boldsymbol{F}_e^i$ describes elastic and $\boldsymbol{F}_{gr}^i$ inelastic, G&R related deformation of the constituent, *i*. The latter is concomitant with the change in shape and volume due to G&R and is not the result of mechanical stresses developed within the tissue.

The stress–strain relationships for the biocomposite during G&R are defined by the elastic potential, Ψ, being the strain energy per volume

$$\Psi = \sum_{i\in I}\Psi^i\left(\boldsymbol{F}_e^i\right) = \sum_{i\in I}\rho_0^i W^i\left(\boldsymbol{F}_e^i\right). \tag{D.5}$$

Here, all quantities are related to an individual constituent, Ψ^i, W^i, ρ_0^i representing the strain energy per unit volume, the strain energy per unit mass and the mass density per unit reference volume, respectively, such that: $\Psi^i(\boldsymbol{F}_e^i) = \rho_0^i W^i(\boldsymbol{F}_e^i)$. This presentation is particularly useful in contexts of G&R, where mass changes significantly over time.

Lightning Source UK Ltd.
Milton Keynes UK
UKHW051627131021
392129UK00003B/68

9 780750 335652